Child-Centered Math

Addition and Subtraction

35 Hands-On Activities
Grades 2–3

Written by Janet Bruno
Illustrated by Terri Sopp Rae
Project Director: Carolea Williams

CTP ©1997, CREATIVE TEACHING PRESS, Inc., Cypress, CA 90630

Reproduction of activities in any manner for use in the classroom and not for commercial sale is permissible. Reproduction of these materials for an entire school or for a school system is strictly prohibited.

Table of Contents

To the Teacher i
Getting Started ii
About Addition and Subtraction iii

Addition Activities

How Many Ways? 1
Egg Carton Shake 2
Building Doubles 3
Doubles Plus One 4
Number Line Game 5
Graphing Sums 6
All Ways 7
Magic Triangles 8
Trading for Tens 9
Dimes and Pennies 10
Roll and Add 11
Addition Race 12
How Much Did You Save? 13
Alphabet Addition 14
Target Number 15

Subtraction Activities

Subtraction Grab Bag 16
Subtraction Grid 17
Heads and Tails Subtraction 18
Card Game 19
Telling Tales 20
Need to Regroup? 21
Ten in a Bag 22
Pick Three 23
Spin and Take Away 24
Partner Subtraction 25
Subtraction Spin 26
Coupon Subtraction 27
Weigh and Subtract 28
Making Problems 29
Spending Ten Dollars 30

Mixed Practice

Addition and Subtraction Toss 31
Calendar Addition and Subtraction 32
Math Riddles 33
Add or Subtract 34
Name the Number 35

Parent Letter 36
Tens and Ones Mat Reproducible 37

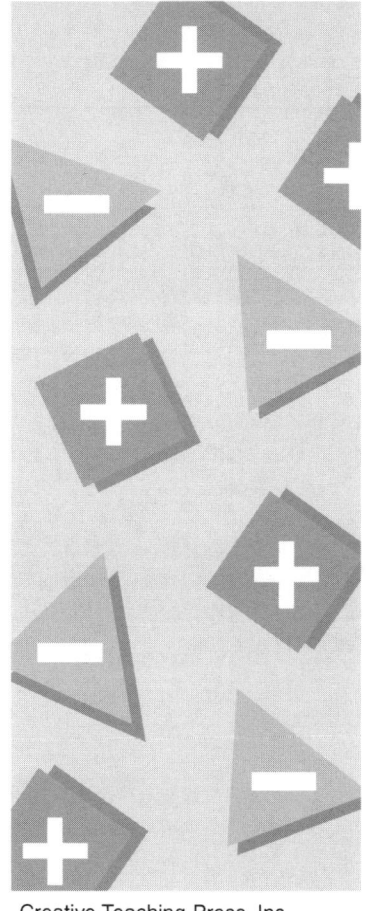

To the Teacher

The child-centered activities in this book are perfect for extending your students' understanding of addition and subtraction. These 35 hands-on activities focus on basic facts to 20 and addition and subtraction with regrouping. There are opportunities for mixed practice as well as separate addition and subtraction activities.

Addition and Subtraction is designed as a handy resource for teachers, not a prescribed continuum of addition and subtraction activities. Integrate the activities into your current mathematics program, keeping in mind the special needs of your students.

Have fun watching your students get excited about math as they participate in these hands-on addition and subtraction activities.

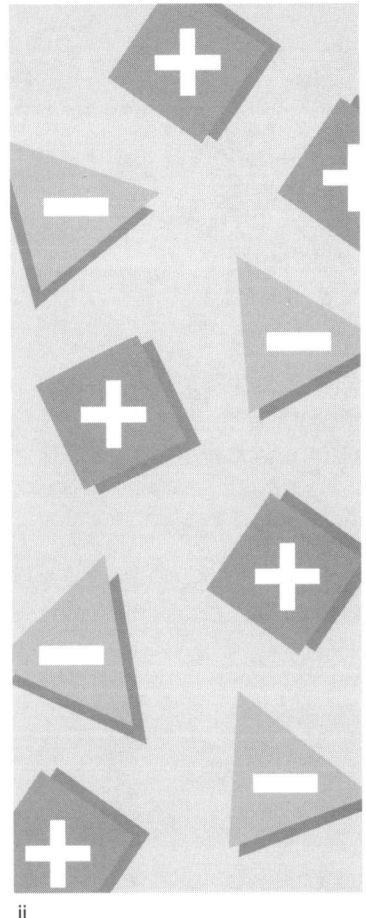

Getting Started

Most activities can be implemented in small- or large-group settings, but some are best suited to learning centers where a few students can work independently. When planning the amount of adult guidance or participation required, keep in mind the materials to be used, the type of work space needed, and the activity level of the project.

Addition and Subtraction is ideal for involving teacher aides or parent volunteers in the classroom. The directions are simple and easy to follow, and students will quickly become engaged in the activities. It may be helpful to place a laminated copy of each activity in a box with the materials. This allows for instant setup and cleanup.

Although most of the activities use a variety of inexpensive, readily-available materials, some parent donations may be helpful. A parent letter is included on page 36 to help you obtain various consumable materials.

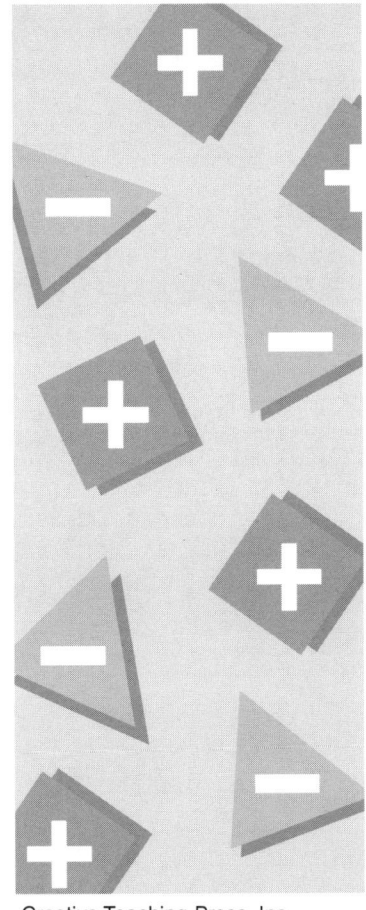

About Addition and Subtraction

Although second- and third-grade students already have a basic understanding of addition and subtraction, it is still important for them to use manipulatives as concepts are extended. Working with manipulatives helps students visualize basic addition and subtraction facts and is a key factor in developing meaning for fact memorization.

Addition and subtraction with regrouping is one of the greatest challenges for second and third graders. Students need plenty of practice with place value models before they attempt to solve problems that involve regrouping. It's best to integrate regrouping problems with non-regrouping problems so students learn to evaluate each problem separately to determine if regrouping is needed.

Throughout these activities, encourage students to verbalize their actions to establish addition and subtraction language patterns. Discuss their findings, especially alternative ways to solve the same problem. As a follow-up, have students describe their experiences and explain strategies in math journals.

How Many Ways?

Activity 1

Materials
- linking cubes (2 colors)
- chalkboard
- chalk

Procedure
1. Have students work with a partner or in small groups.
2. Call out a number from 10 to 20.
3. Ask students to use the cubes to show different addition combinations for that number. For example, for the sum of 15, students can link nine red cubes and six blue cubes, and seven red cubes and eight blue cubes.
4. As students share their addition combinations, write the corresponding equations on the chalkboard.
5. Call out a new number and repeat the activity.

Addition and Subtraction

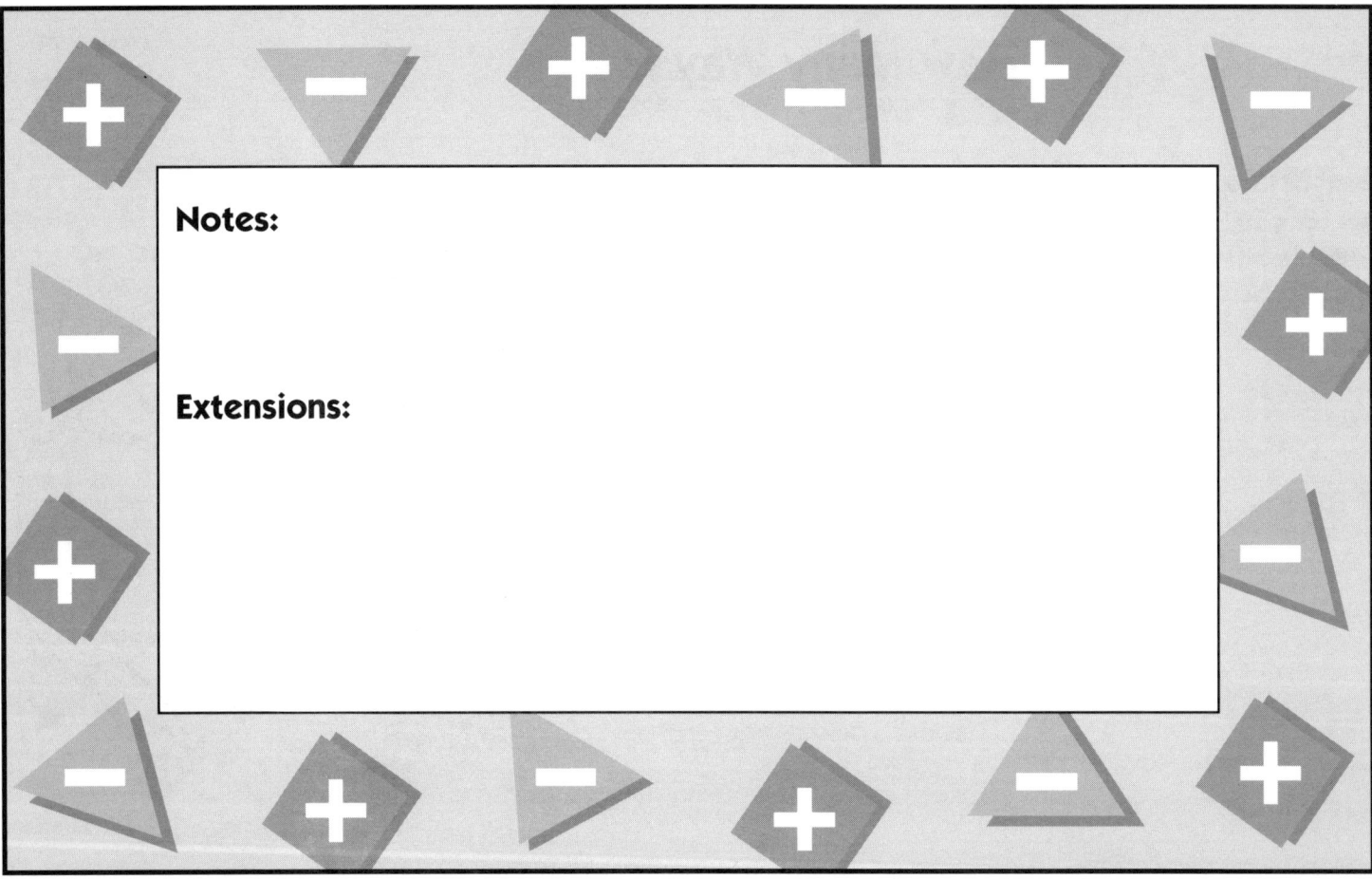

Notes:

Extensions:

Egg Carton Shake

Activity 2

Materials

- markers
- egg cartons
- beans
- paper
- pencils

Procedure

1. Have each pair of students use a marker to write numbers *4–9* inside the egg carton cups. Each number should be written twice.
2. Have Player A place two beans in the carton, close the lid, and shake.
3. Tell Player B to open the lid and write the addition problem. For example, if the beans land in cups 6 and 9, the student writes *6 + 9 = ___*.
4. Instruct both students to calculate the answer, using counters as necessary.
5. Ask Player B to record the answer.
6. Have students continue the game, taking turns shaking and recording.

Extension Have students work with larger numbers to practice addition with regrouping.

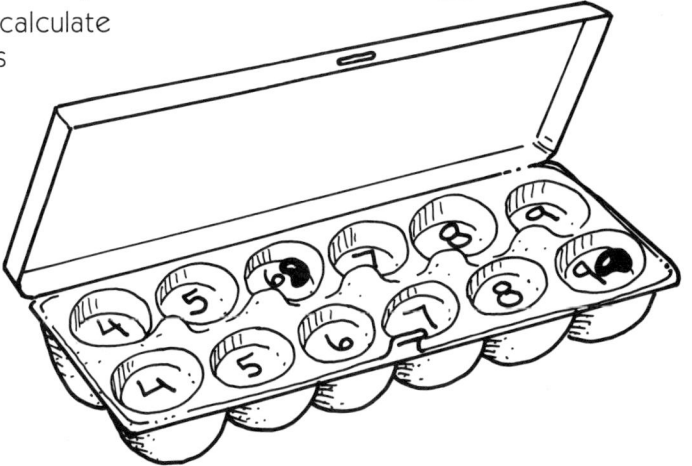

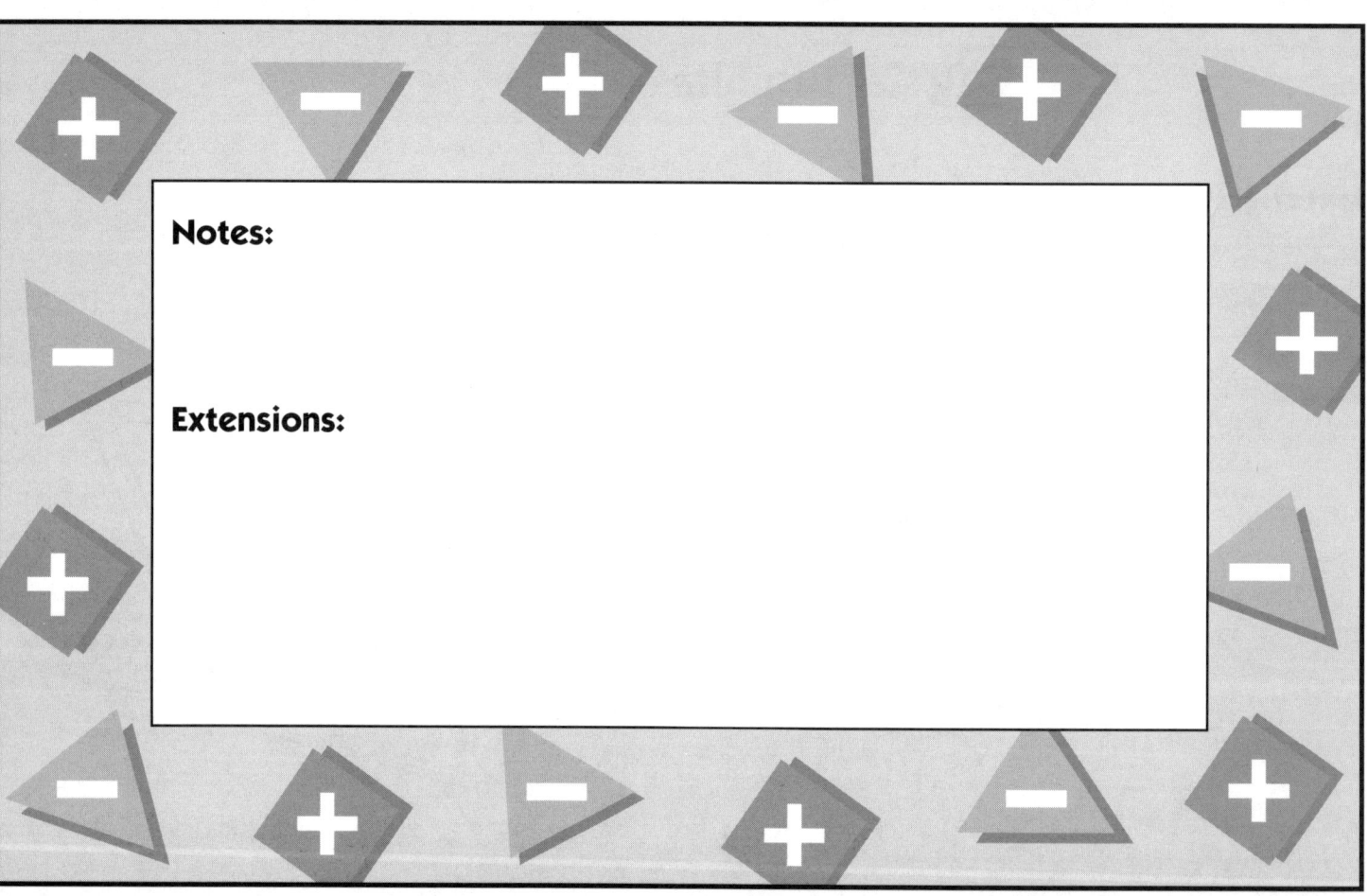

Notes:

Extensions:

Building Doubles

Activity 3

Materials
- linking cubes
- paper
- pencils

Procedure
1. Have students work with a partner. Give each child ten linking cubes.
2. Ask one student to build a train one to ten cubes long.
3. Tell his or her partner to double the length of the train using another color.
4. Ask students to record the addition equation.
5. Have students repeat steps 1–3 until they have recorded all possible doubles problems.

Addition and Subtraction

Notes:

Extensions:

Doubles Plus One

Activity 4

Materials
- 1" graph paper
- scissors
- counters
- chalkboard
- chalk

Procedure
1. Cut the graph paper into 2" x 9" strips. Give each student 20 counters and one strip.
2. Write a doubles fact such as *8 + 8* on the chalkboard.
3. Have students use counters to show the fact on the graph paper strip.
4. Have students add one more counter to show a doubles-plus-one fact.
5. Record the doubles-plus-one problem on the board.
6. Continue with similar problems.

Variation Have students work independently, recording doubles problems and doubles-plus-one problems on paper.

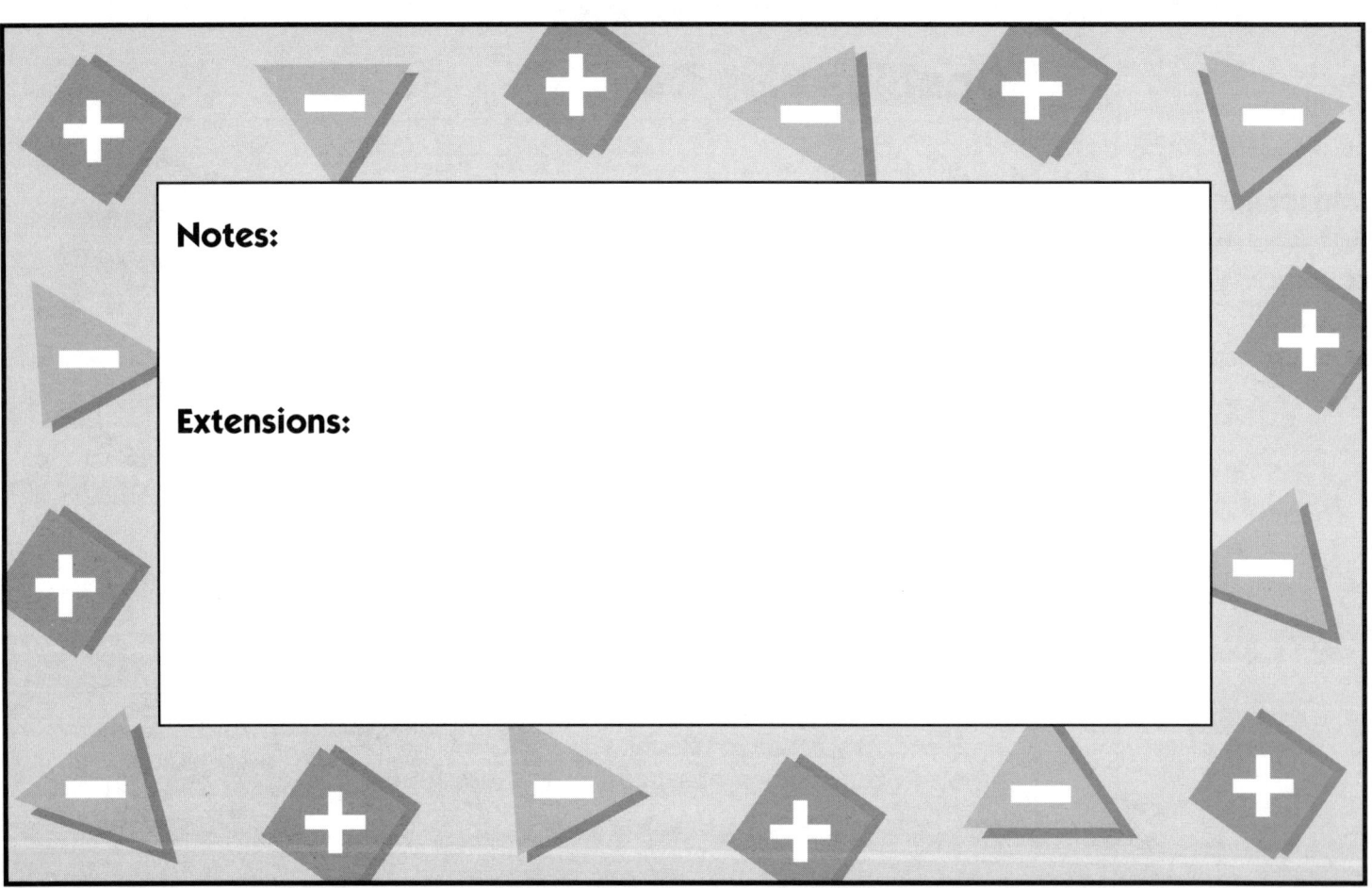

Notes:

Extensions:

Number Line Game

Activity 5

Materials

- 1" graph paper
- scissors
- tape
- markers
- 2-color counters
- paper
- pencils

Procedure

1. Help students prepare number lines from 1–18 using graph paper, scissors, tape, and markers.
2. Name a sum, and have students find different addition facts for that number. Have them use different-colored counters for the two addends.
3. Ask students to record each addition fact on paper.
4. When students have discovered all possible facts for one sum, name another sum and repeat steps 2 and 3.

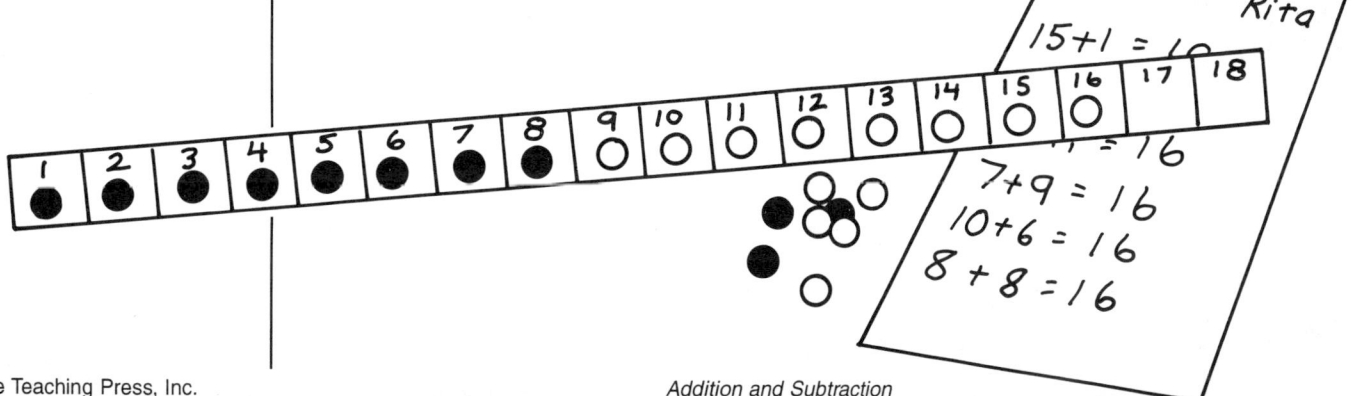

Addition and Subtraction

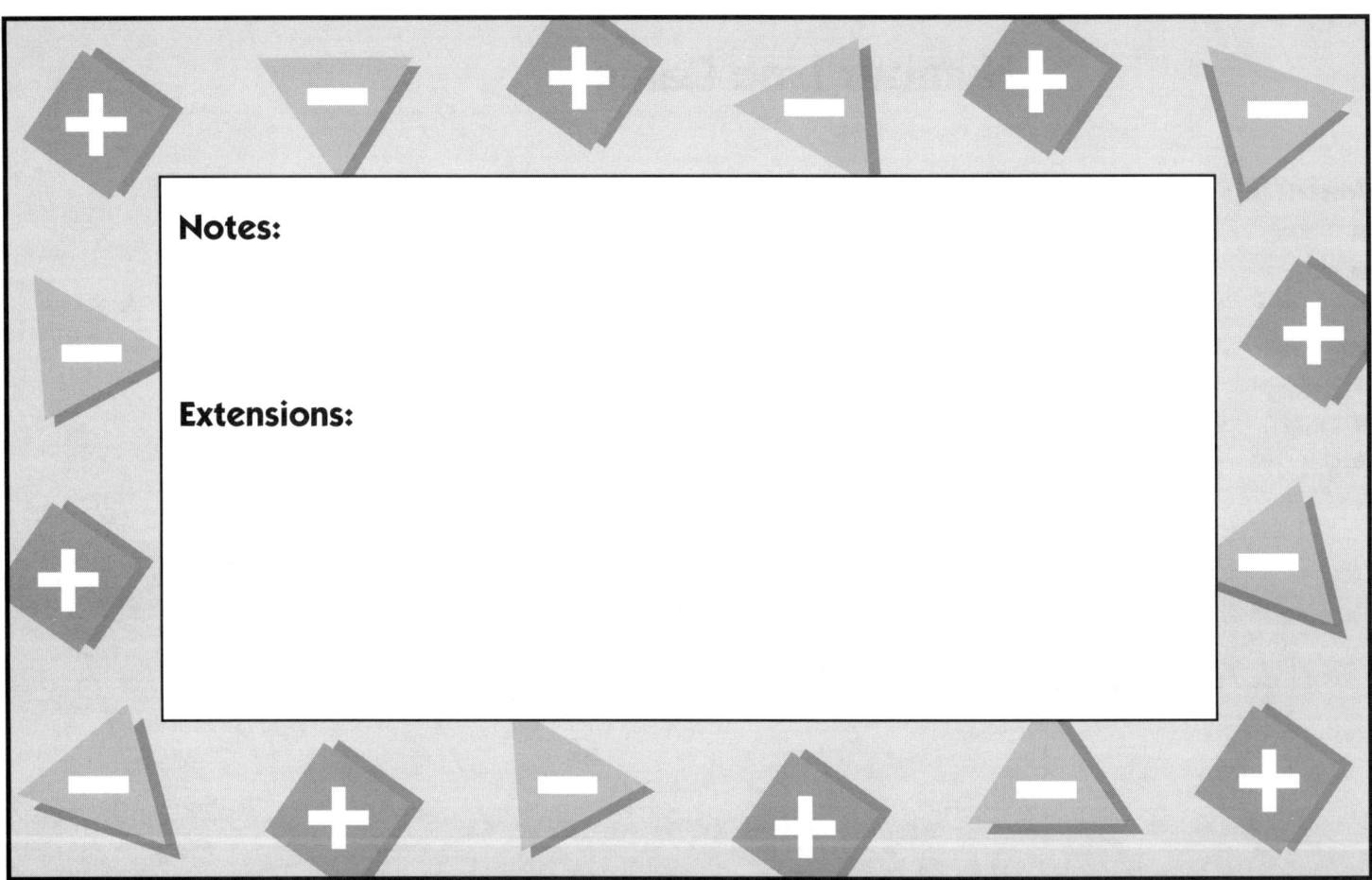

Notes:

Extensions:

Graphing Sums

Activity 6

Materials
- small blank cubes
- 1-cm graph paper
- crayons or markers
- pencils

Procedure
1. Write the numbers 5–10 on each blank cube, one number on each cube face. (Or, use small self-adhesive labels on dice, with numbers written on them.)
2. Have students work in pairs. Give each pair two number cubes and a piece of graph paper.
3. Help students prepare a simple graph showing sums 10–20 on the horizontal axis and numbers 1–10 on the vertical axis.
4. Ask students to take turns tossing the cubes, finding the sum of the two numbers, and coloring in a box above the correct sum on the graph.
5. After 30 rolls are recorded, ask students to compare their results with classmates. Do certain sums appear more often?

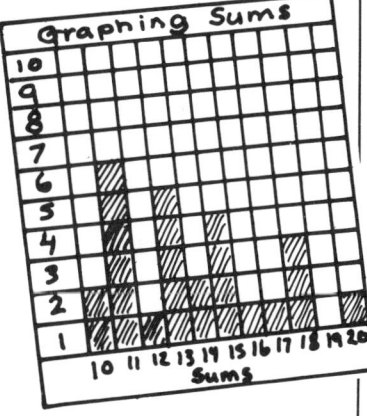

Addition and Subtraction

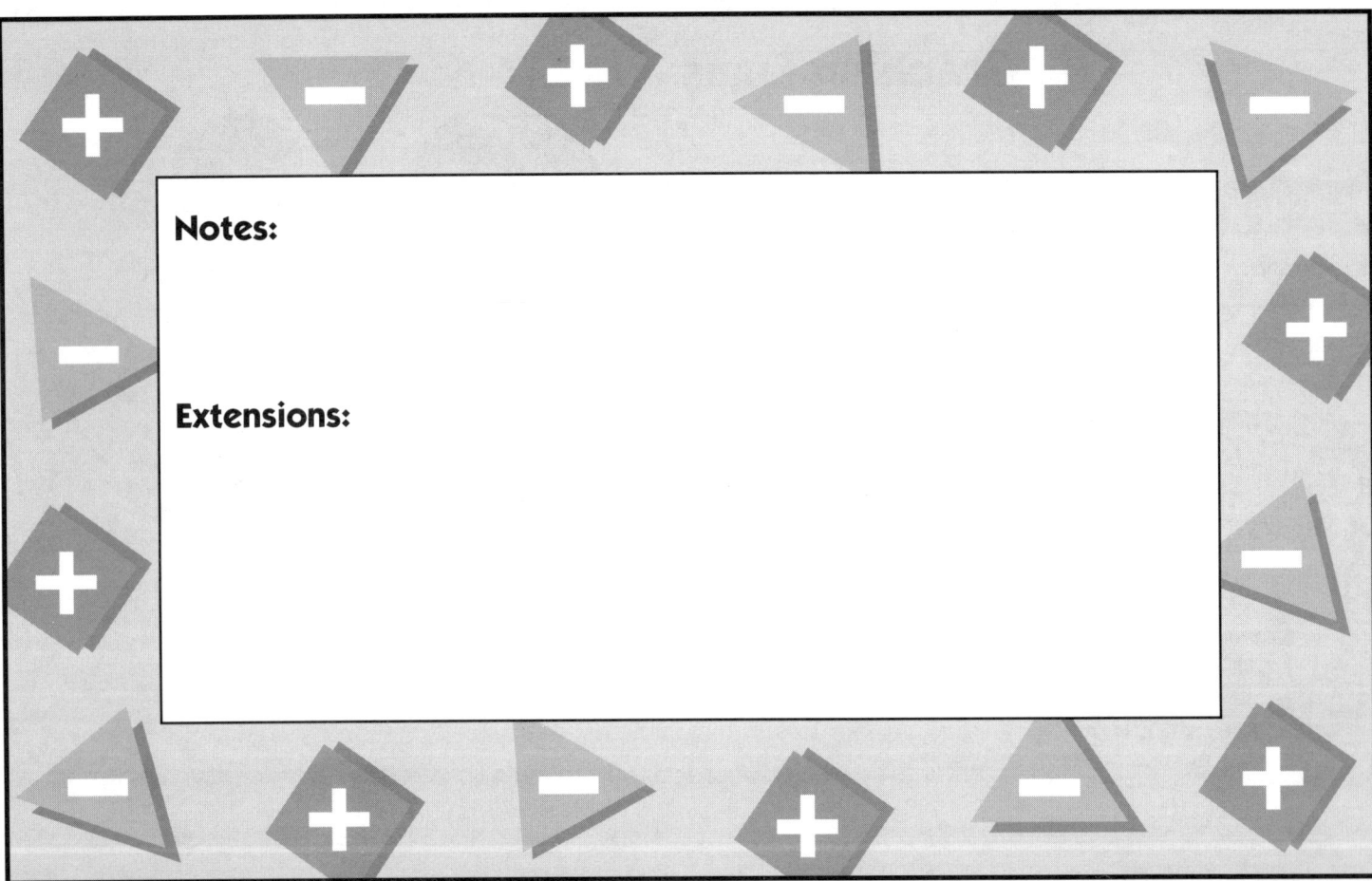

Notes:

Extensions:

All Ways

Activity 7

Materials

- counters
- 12" x 18" construction paper (folded in thirds)
- paper
- pencils

Procedure

1. Ask students to work in pairs. Give each pair 20 counters and a piece of construction paper.
2. Call out a number from 10 to 20 (e.g., 15). Have one student from each pair divide 15 counters into three groups on construction paper, forming an addition problem with three addends.
3. Ask his or her partner to write the matching addition equation.
4. Challenge students to find as many three-addend problems as they can for the same sum.
5. Call out a new number and have partners switch roles.

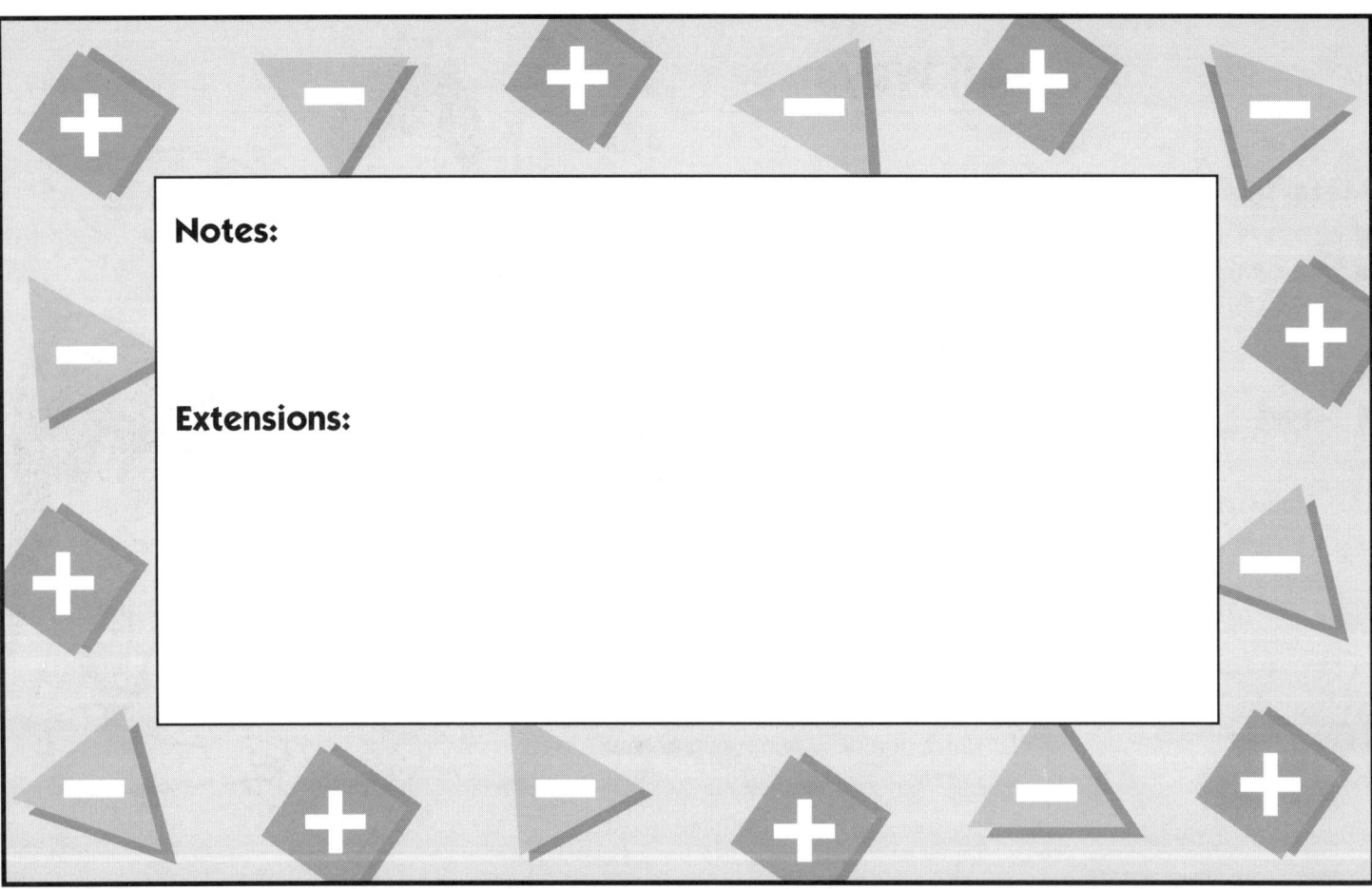

Notes:

Extensions:

Magic Triangles

Activity 8

Materials

- 20–30 small paper plates, paper circles, or game chips (labeled 0–9)
- markers
- 8 index cards (labeled with sums 11–18)

Procedure

1. Place materials at a center.
2. Tell students to select an index card and a set of paper plates numbered 0–9.
3. Have them use six of the plates to build a triangle. Each side of the triangle must match the sum on the index card.
4. Let students then choose a new card and continue as above.

Extension Challenge students to build as many different "magic triangles" as they can for each sum.

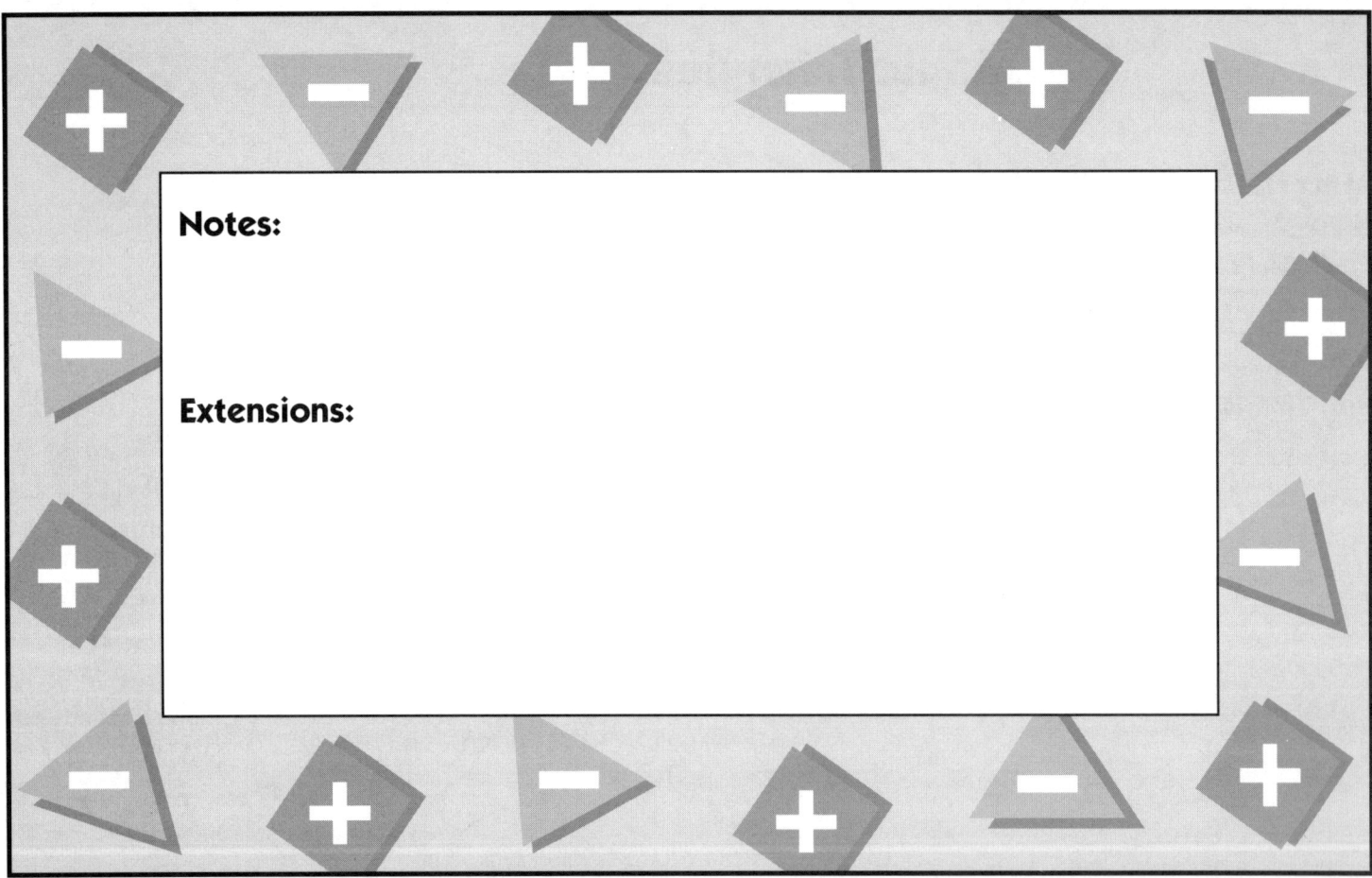

Notes:

Extensions:

Trading for Tens

Activity 9

Materials
- index cards
- markers
- tens and ones manipulatives (Base Ten Blocks, beansticks and beans, linking cubes)

Procedure
1. Write addition problems on the index cards, using a two-digit and one-digit number. Place the cards and tens and ones manipulatives at a center.
2. Ask students to work in pairs. Have one student model the problem with the tens and ones manipulatives.
3. His or her partner then trades ten ones for a ten, if possible, and tells the total number.
4. Have students continue with the activity, switching roles.

Variation Focus on subtraction with regrouping instead of addition.

Addition and Subtraction

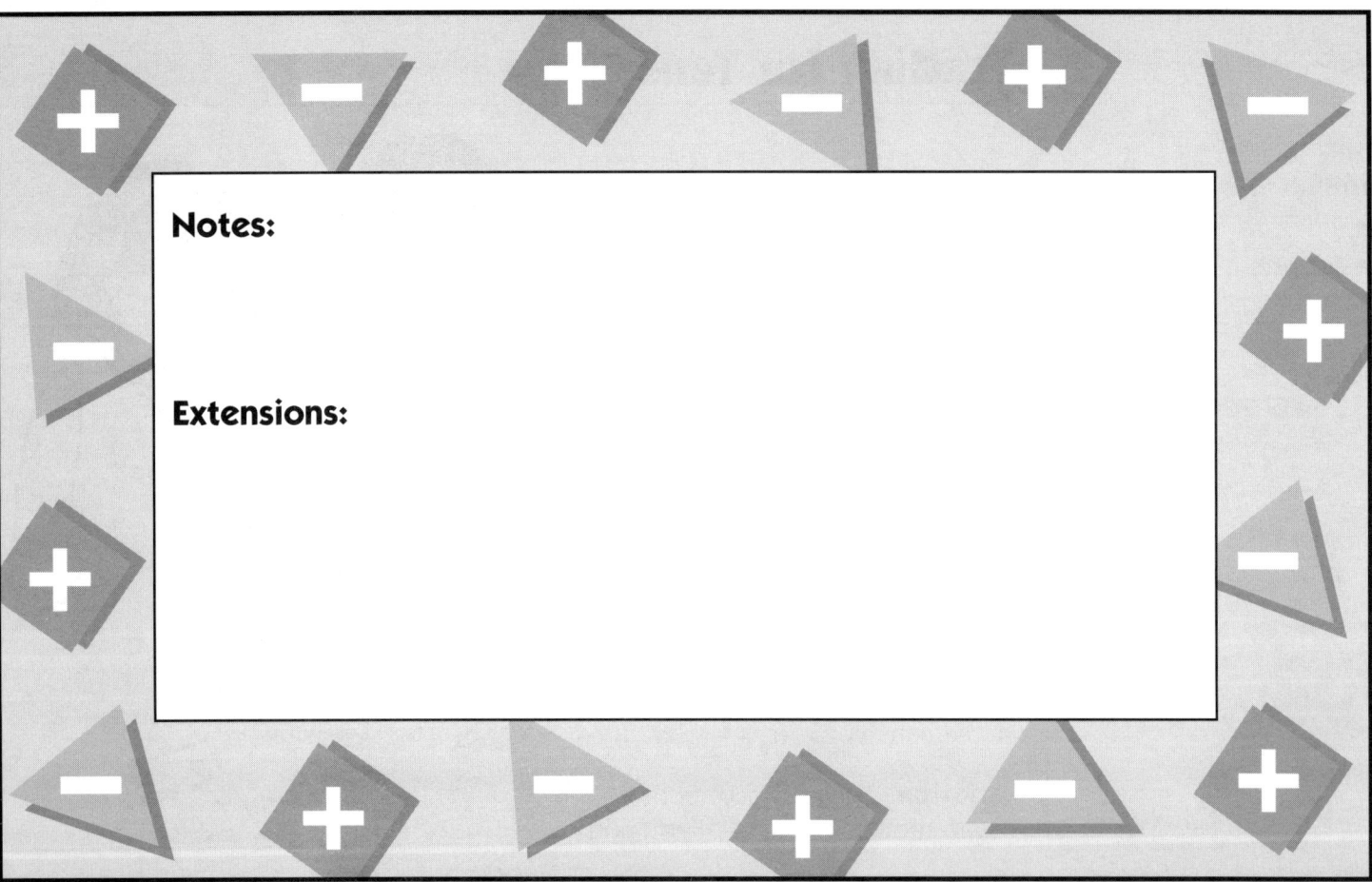

Notes:

Extensions:

Dimes and Pennies

Activity 10

Materials
- dimes
- pennies
- Tens and Ones Mat reproducible (page 37)
- chalkboard
- chalk

Procedure
1. Working with a small group, give each student 20 dimes, 20 pennies, and a Tens and Ones Mat.
2. Write a two-digit number on the board, and ask students to represent the number with dimes and pennies on the mat.
3. Repeat step 2 with a second number.
4. Have students add the dimes and pennies together, exchanging ten pennies for a dime when necessary.

Variation Adapt the same activity for subtraction.

Extension Use three-digit numbers and include dollars in the materials.

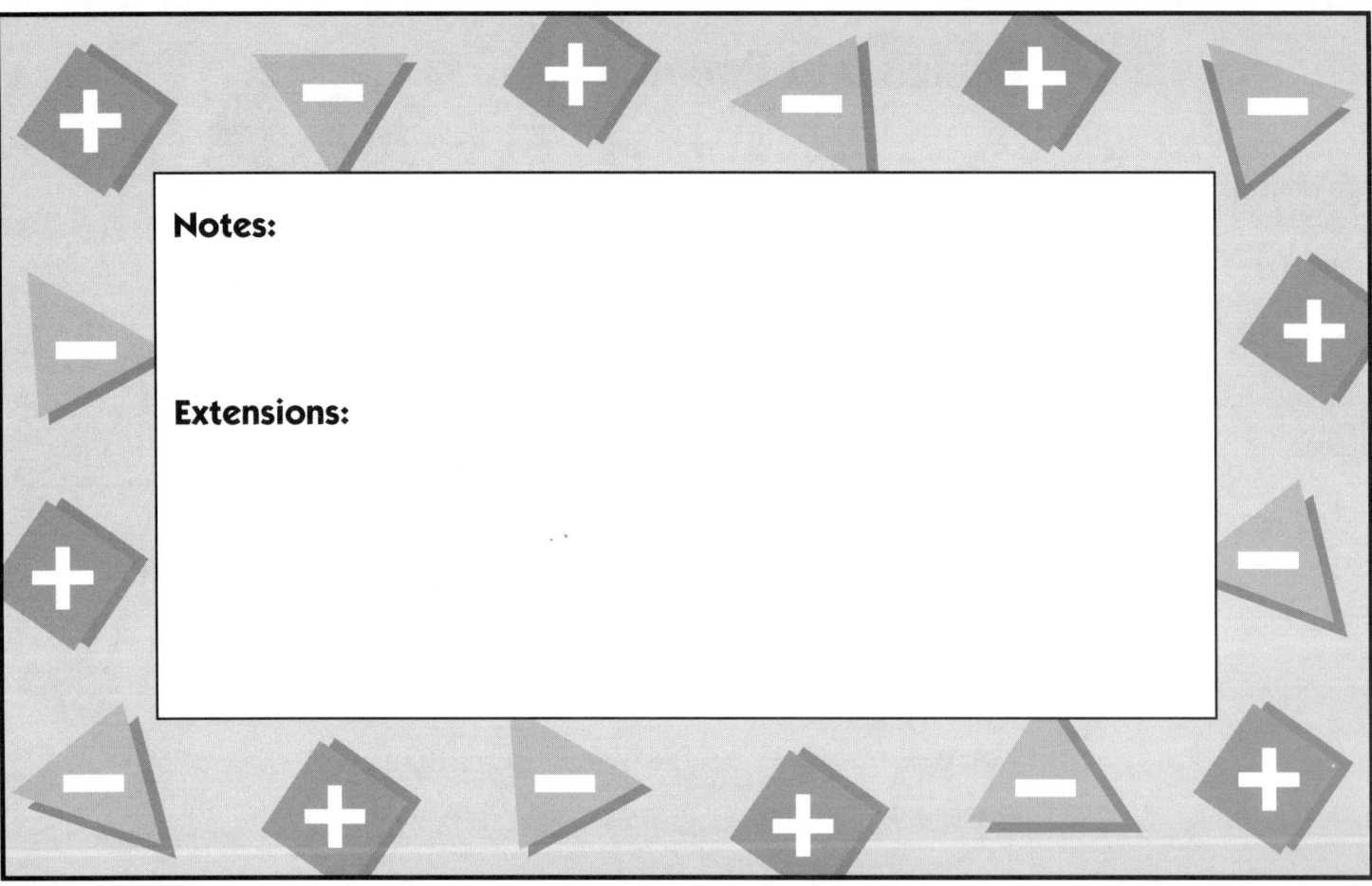

Notes:

Extensions:

Roll and Add

Activity 11

Materials

- dice (2 colors)
- tens and ones manipulatives (Base Ten Blocks, beansticks and beans, linking cubes)
- paper
- pencils

Procedure

1. Designate which color die will represent tens and which will be ones.
2. Assign partners and have one student from each pair roll the dice (one of each color) and record the two-digit number on paper.
3. Have his or her partner repeat step 2 for the second addend.
4. Ask both students to use tens and ones manipulatives to solve the problem and record the answer.
5. Have students continue playing as described.

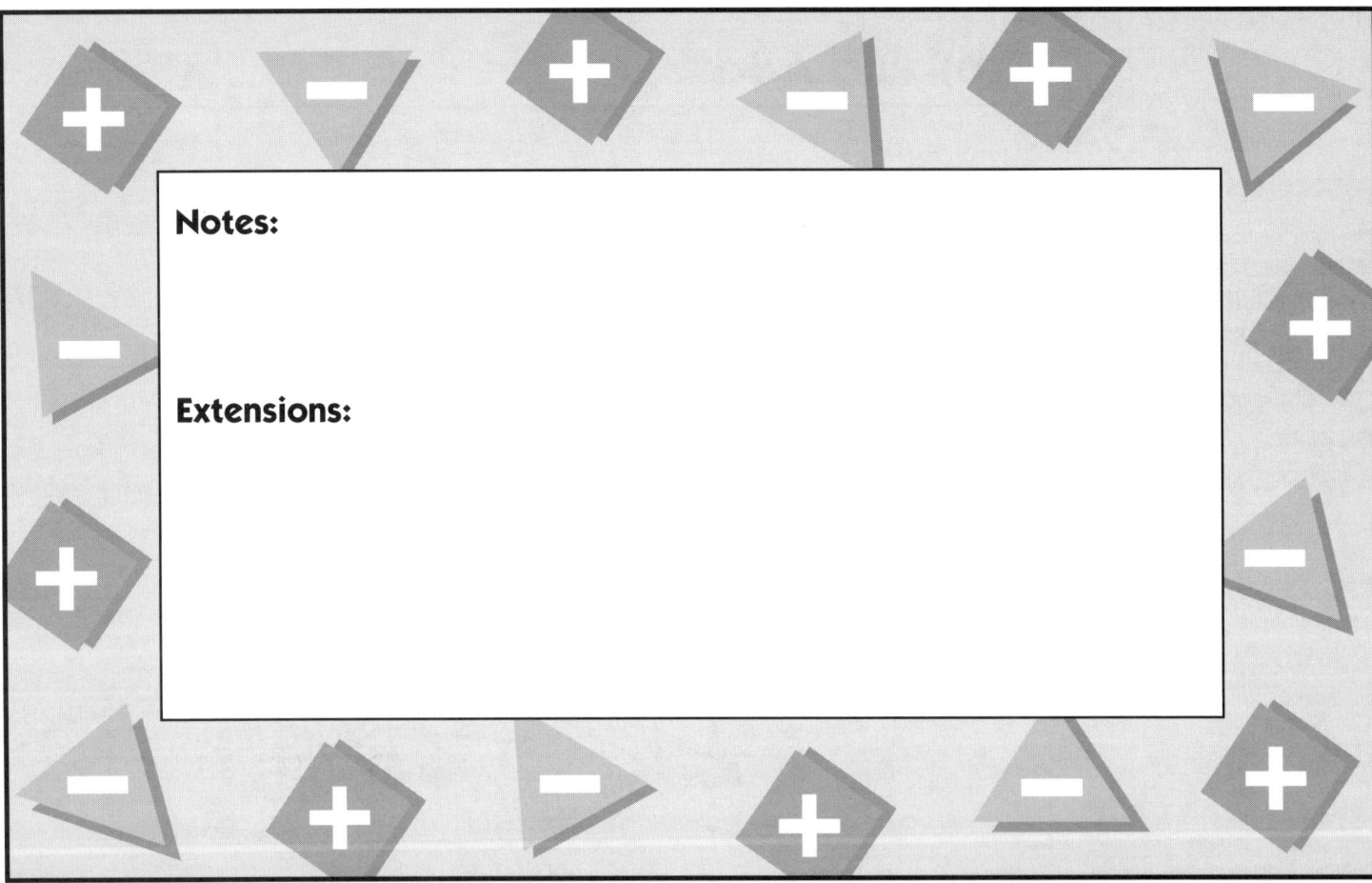

Notes:

Extensions:

Addition Race

Activity 12

Materials
- dice
- tens and ones manipulatives (Base Ten Blocks, beansticks and beans, linking cubes)
- paper
- pencils

Procedure
1. Divide the class into small groups or partners. Give each group a die.
2. Have each player roll the die and record the number.
3. On the second roll, have players add the new number to the first number. Invite students to use manipulatives to calculate answers.
4. Ask students to take turns rolling and adding until one player reaches or exceeds 50.

Extension Have students play with two dice. The first to reach or exceed 100 is the winner.

Addition and Subtraction

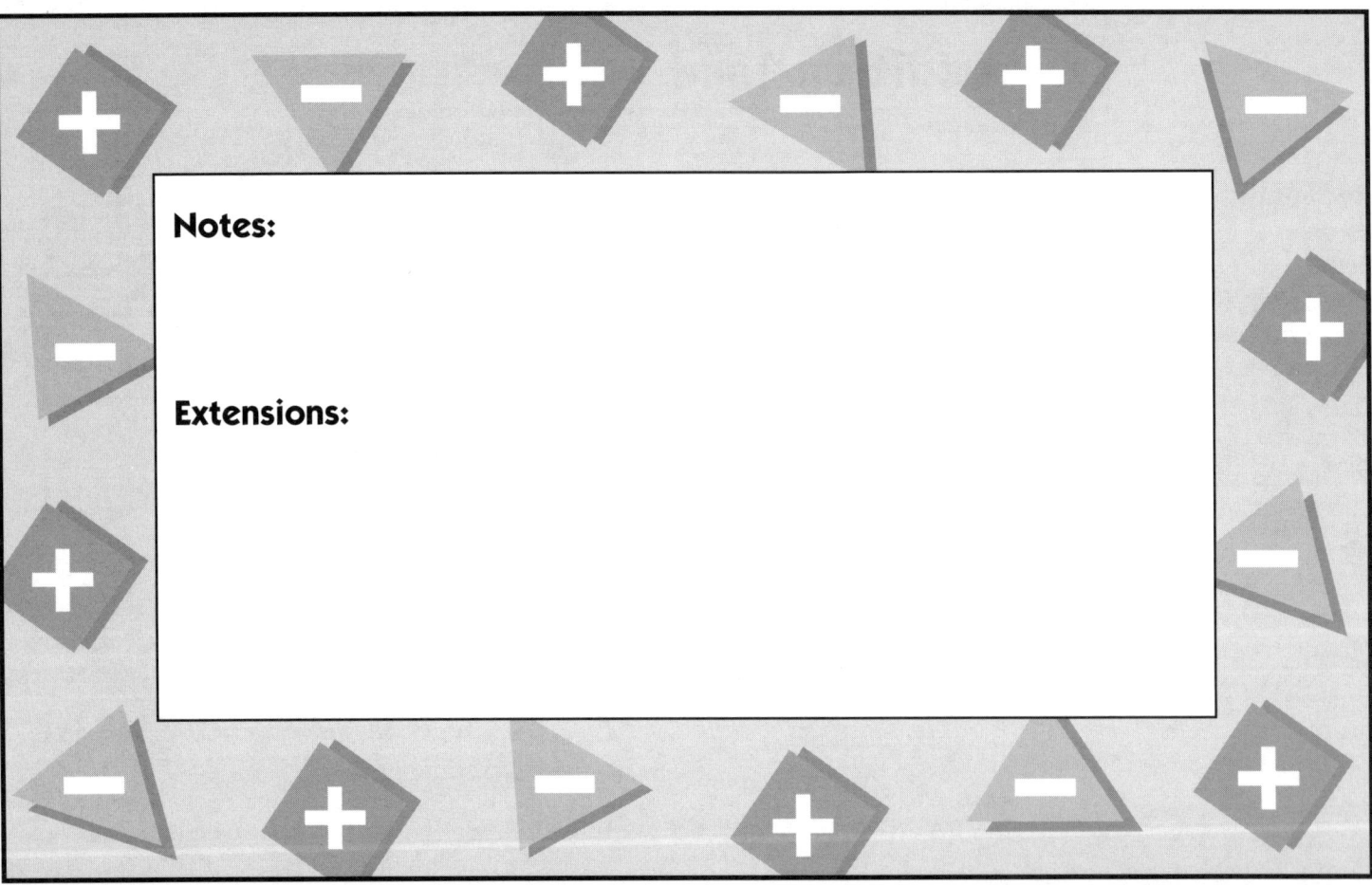

Notes:

Extensions:

How Much Did You Save?

Activity 13

Materials

- newspaper and magazine coupons
- laminating materials or tagboard and glue
- paper
- pencils
- play coins

Procedure

1. Laminate the coupons or glue them on tagboard.
2. Place coupons face down in the middle of a small group of students.
3. One at a time, have each student choose a coupon and record the value.
4. As students select a second coupon, ask them to use coins to add up the savings and record the addition.
5. Have students take turns picking coupons until one player reaches $3.00 or more in savings.

Addition and Subtraction

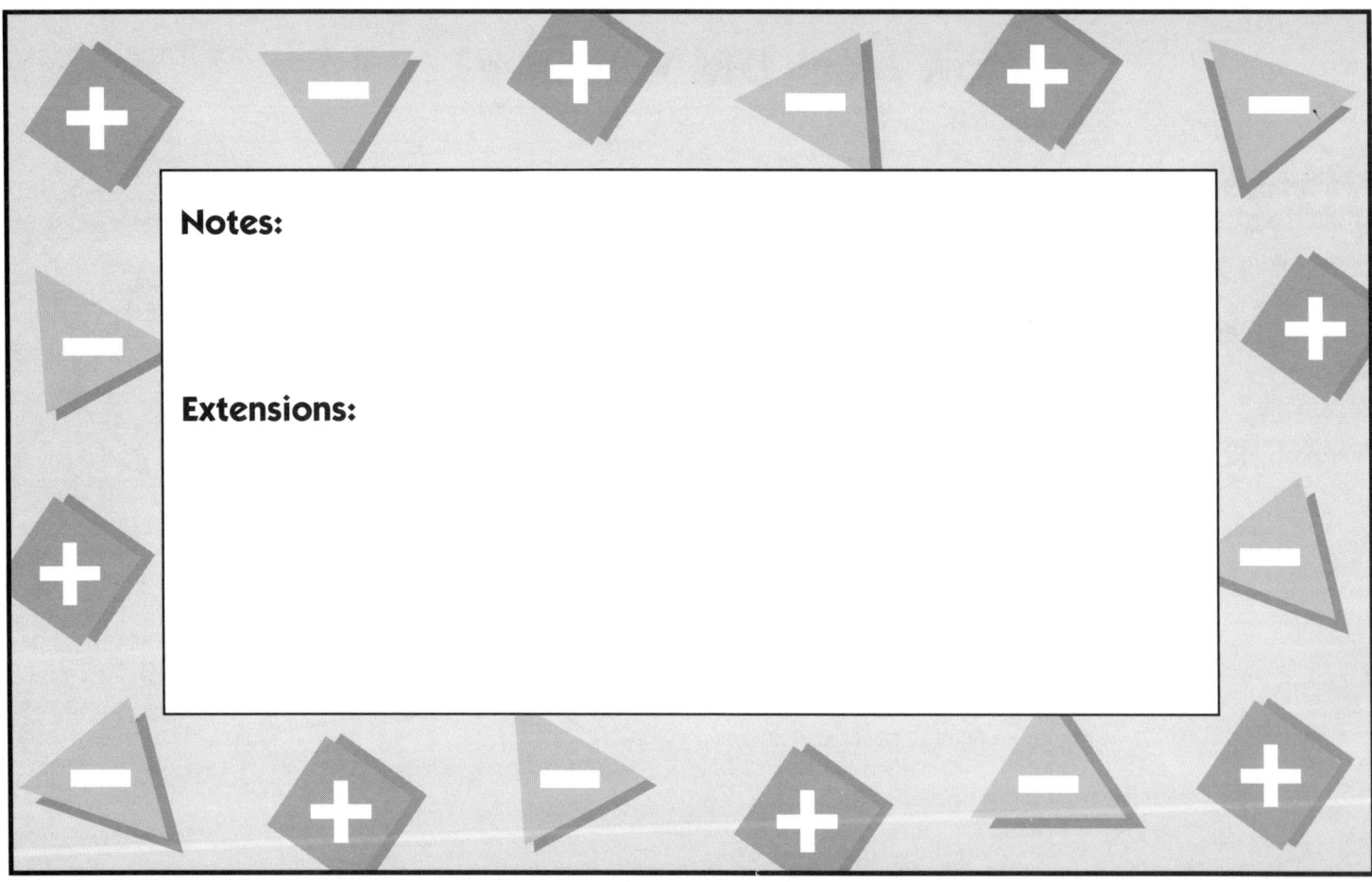

Notes:

Extensions:

Alphabet Addition

Activity 14

Materials

- chalkboard
- chalk
- paper
- pencils

Procedure

1. Write the alphabet on the chalkboard and assign one number (1–26) to each letter (A = 1, B = 2, C = 3, . . . Z = 26).
2. Write a word on the board. Ask students to add up the value of the word. (For example, the value of *crayon* would be 76.)
3. Let students name words to spell and add.

Extension   Challenge students to find words with a particular value. (For example, *jet, ant,* and *tan* all have a value of 35.)

crayon
3 + 18 + 1 + 25 + 15 + 14 = 76

Addition and Subtraction

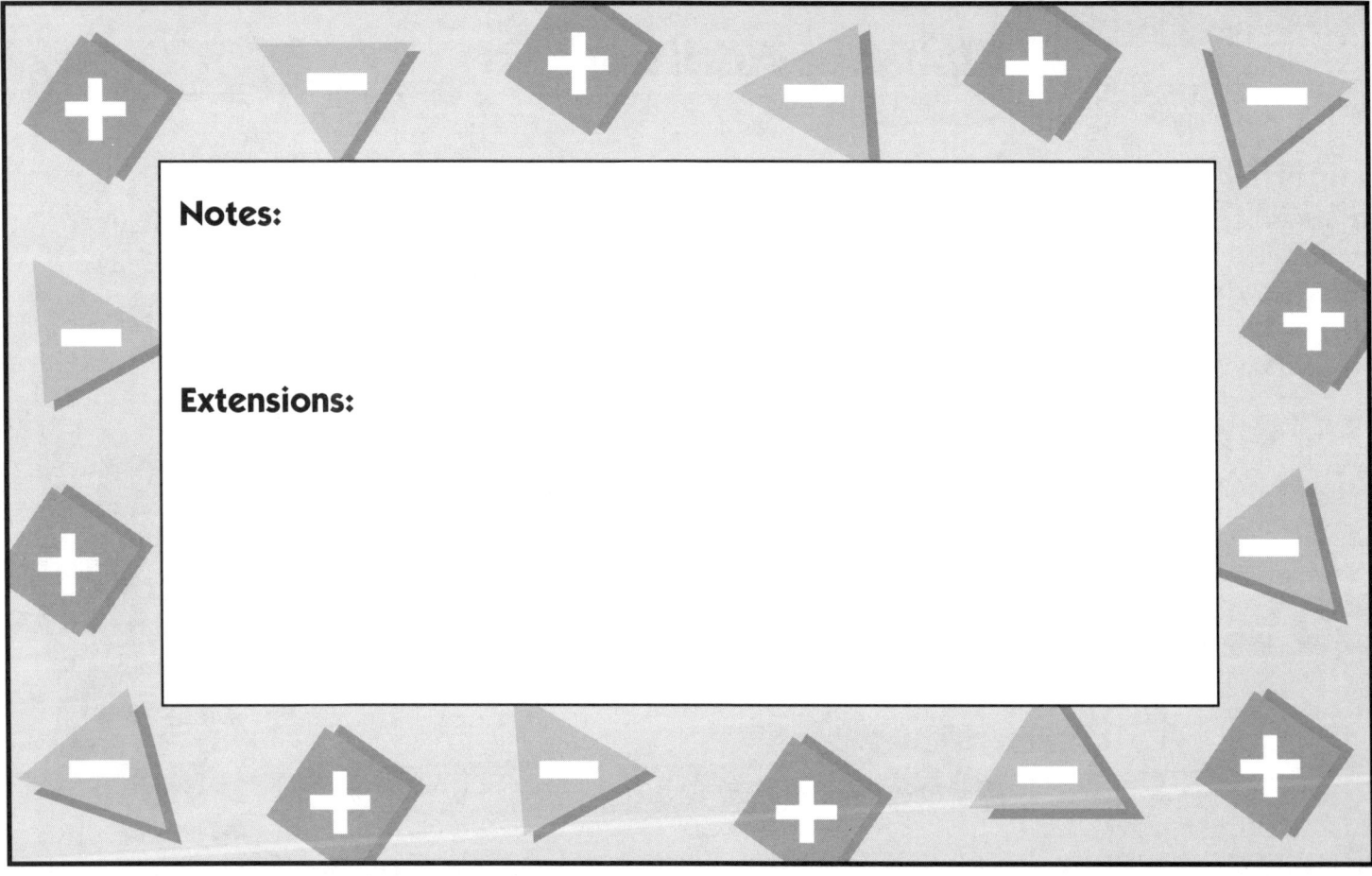

Notes:

Extensions:

Target Number

Activity 15

Materials

- number card sets (labeled 0–9)
- tens and ones manipulatives (Base Ten Blocks, beansticks and beans, linking cubes)
- paper
- pencils

Procedure

1. Divide the class into small groups and give two sets of number cards to each group.
2. Have students mix up the cards and place them in a pile, face down.
3. Ask each student to write 20 at the top of his or her paper.
4. Designate a target number such as 75 or 100. Ask one student from each group to choose a card and add that number to 20, using manipulatives as needed.
5. Have other players take turns recording the new sum on every turn.
6. The first player to reach or pass the target number scores a point.
7. Have students play additional rounds. The first player to get five points is the winner.

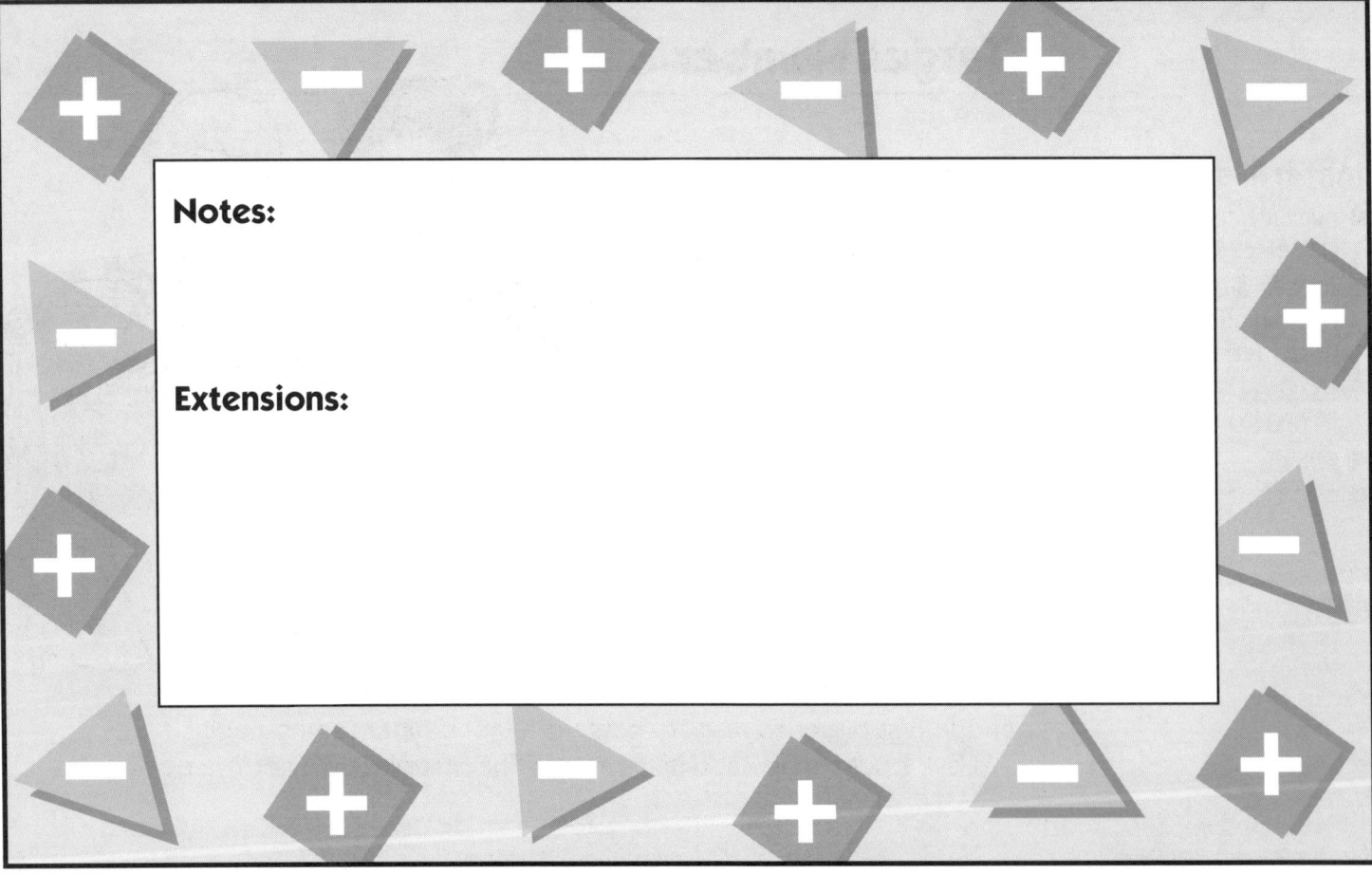

Notes:

Extensions:

Subtraction Grab Bag

Activity 16

Materials
- paper lunch bags
- counters
- paper
- pencils

Procedure
1. Give each student pair a paper bag and some counters. Designate a specific number of counters to put in the bag. For example, if you want students to work on subtraction facts for 15, have them put 15 counters in the bag.
2. Ask one student from each pair to take some counters out of the bag and count them.
3. Without looking, have his or her partner tell how many counters are left in the bag.
4. Have both students look inside the bag to verify the number, and ask one child to write the subtraction equation on paper.
5. Ask students to reverse roles and continue the activity.

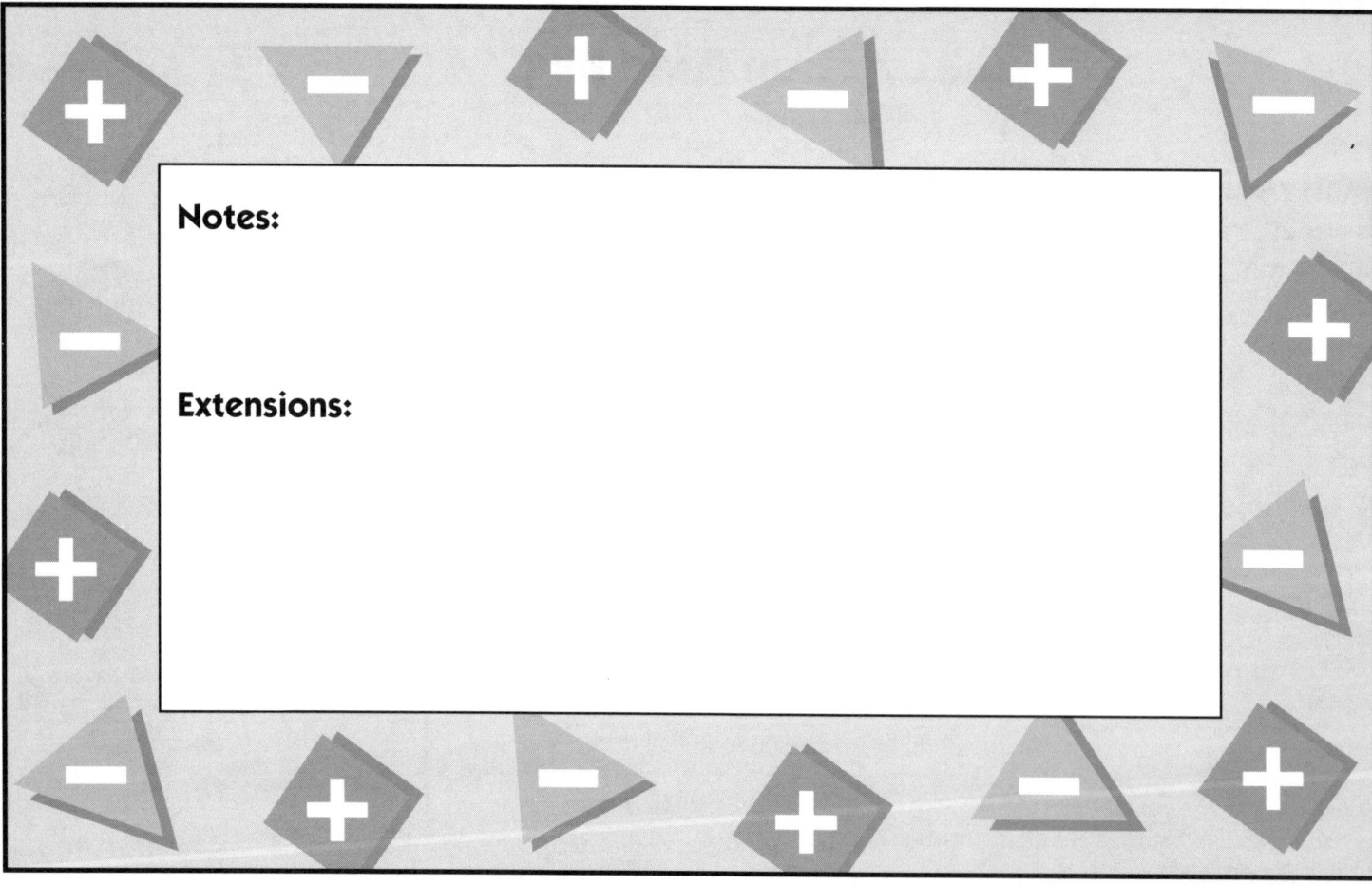

Notes:

Extensions:

Subtraction Grid

Activity 17

Materials
- beans
- grids (2 rows of 10 squares)
- paper
- pencils

Procedure
1. Divide the class into partners, and give each student pair a grid and 20 beans.
2. Have one student in each pair place 10–20 beans on the grid, one bean in each square.
3. Ask his or her partner to remove a number of beans.
4. Have both students verbalize the corresponding subtraction equation and record the equation on paper.
5. Ask students to repeat the activity, taking turns placing the beans on the grid and taking some away.

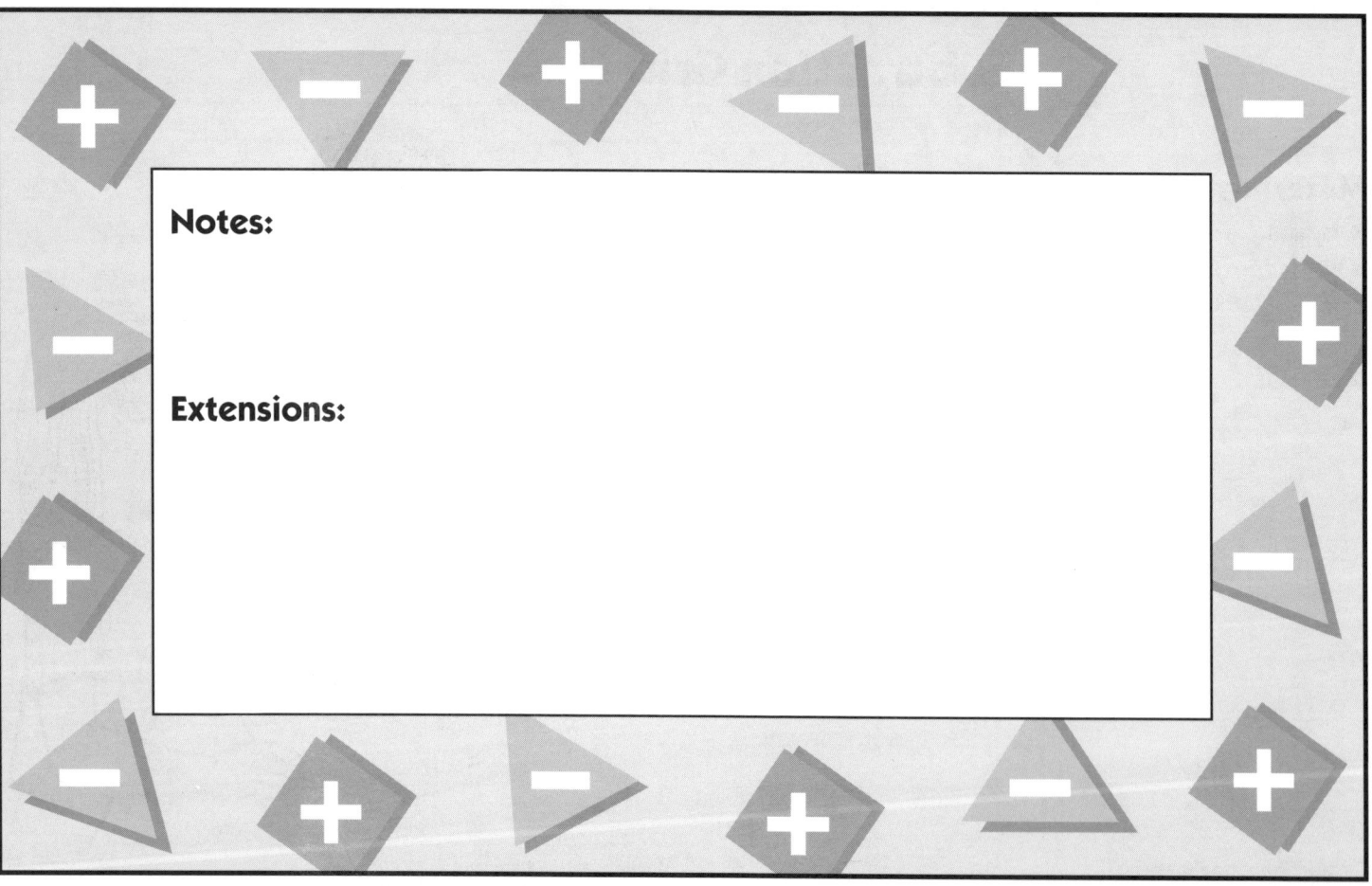

Notes:

Extensions:

Heads and Tails Subtraction

Activity 18

Materials
- pennies
- plastic cups
- paper
- pencils

Procedure
1. Have students work with a partner. Give each student pair 20 pennies and a plastic cup.
2. Designate a specific number of pennies to place in the cup. For example, if you are working on subtraction facts for 14, have students place 14 pennies in the cup.
3. Have Player A shake the cup and spill out the pennies.
4. Ask Player A to count the coins that show heads, while Player B counts the coins that show tails.
5. Have both players write a subtraction equation to match the toss. For example, if nine pennies land heads up, Player A writes *14 - 9 = 5*, and Player B writes *14 - 5 = 9*.
6. Invite students to compare their equations and continue taking turns tossing the pennies.
7. On another day, have students work on facts using a different target number.

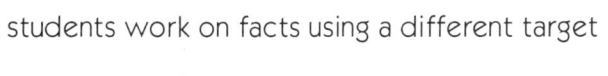

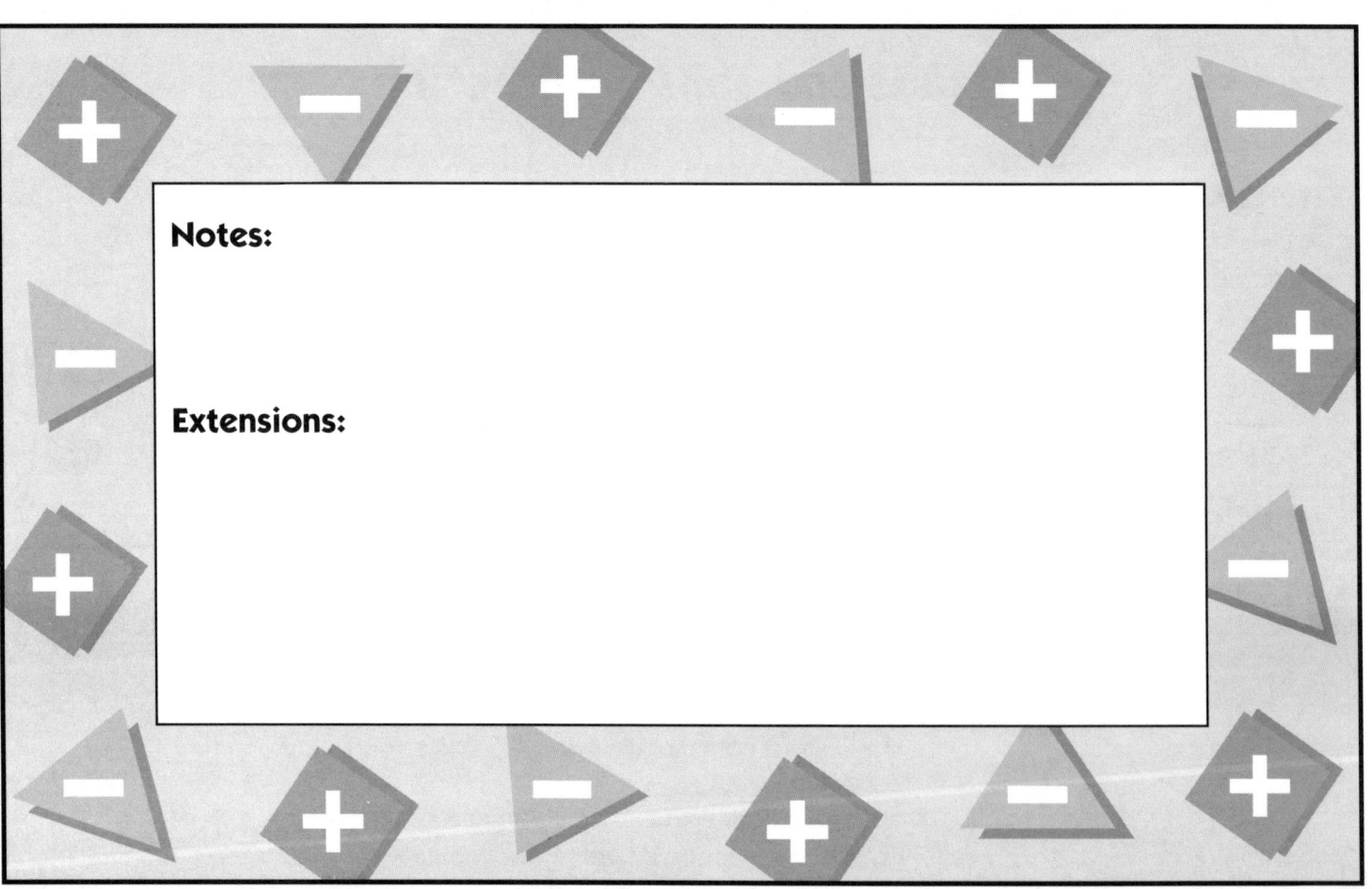

Notes:

Extensions:

Card Game

Activity 19

Materials
- index cards
- markers
- counters
- paper
- pencils

Procedure
1. Number several sets of index cards *5–20*.
2. Divide the class into partners and give each pair two sets of cards and about 40 counters.
3. Have students shuffle all 32 cards, divide them evenly, and place them face down in two piles.
4. Each student draws a card and shows it to his or her partner.
5. Whoever has the greater number computes the difference between the two numbers and takes the corresponding number of counters.
6. When all the cards have been drawn, the player with the most counters wins.

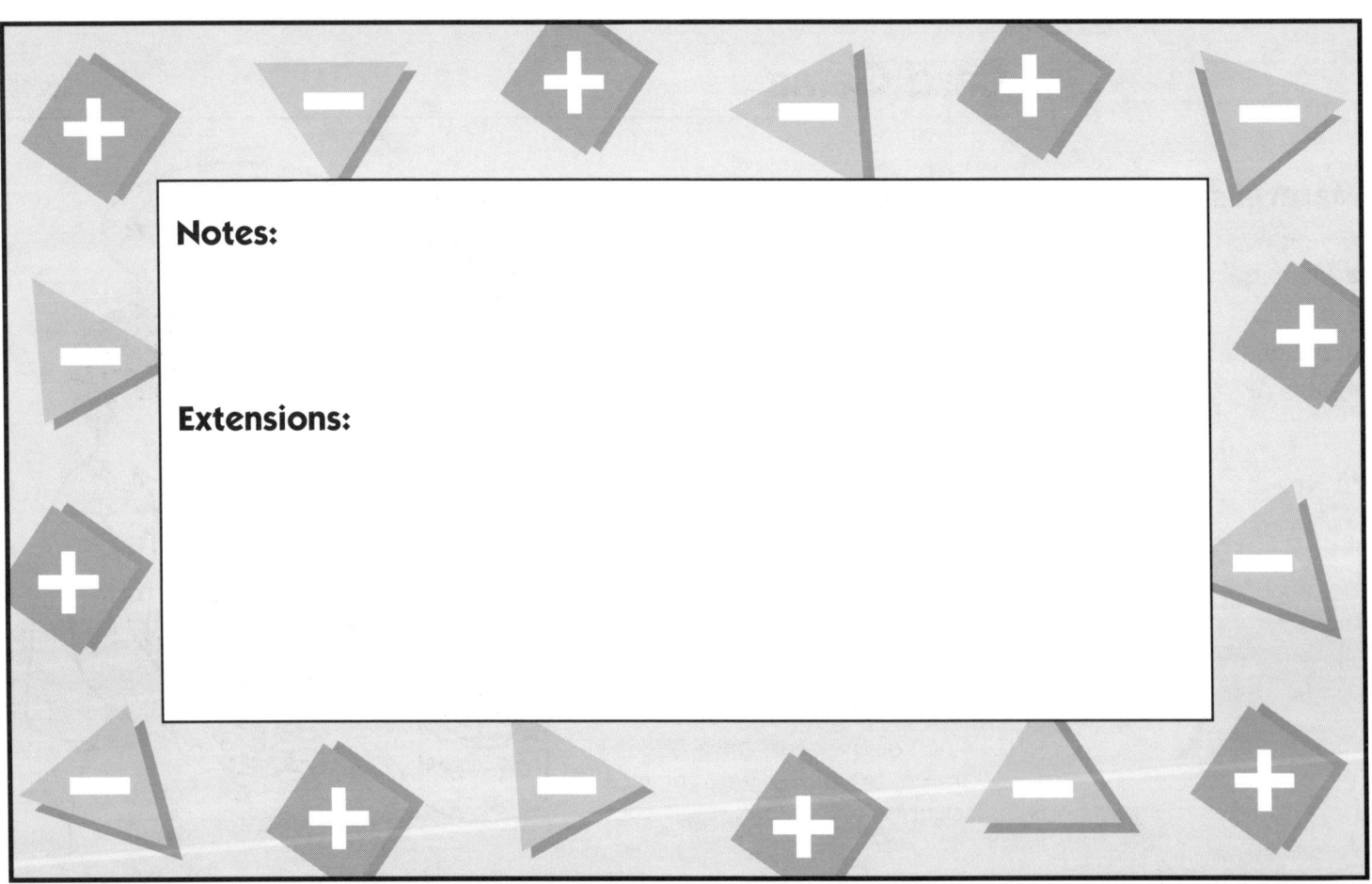

Notes:

Extensions:

Telling Tales

Activity 20

Materials
- tape recorder
- cassette tape
- paper
- pencils
- counters

Procedure
1. Pre-record 10–15 subtraction story problems on a cassette tape.
2. Place materials at a center where two or three students can work together.
3. Have students listen to a problem, use counters to solve the problem, and write down the corresponding subtraction equation.
4. Ask students to compare their work. If they agree, they can move on to a new problem.

Variation Ask students to draw a picture showing the subtraction problem.

Extension Let students record some of their own subtraction story problems.

Notes:

Extensions:

Need to Regroup?

Activity 21

Materials
- 3" x 5" index cards
- markers
- chalkboard
- chalk

Procedure
1. Give each student two index cards and have him or her write *yes* on one card and *no* on the other.
2. Have a volunteer write a subtraction problem on the board and ask, "Do I need to regroup?"
3. The rest of the group indicate their answers by holding up *yes* or *no* cards. Ask for a volunteer to explain why.
4. Have students continue taking turns at the board.

Variation Have a volunteer solve the problem at the chalkboard and ask the class to verify the answer with their *yes* and *no* cards.

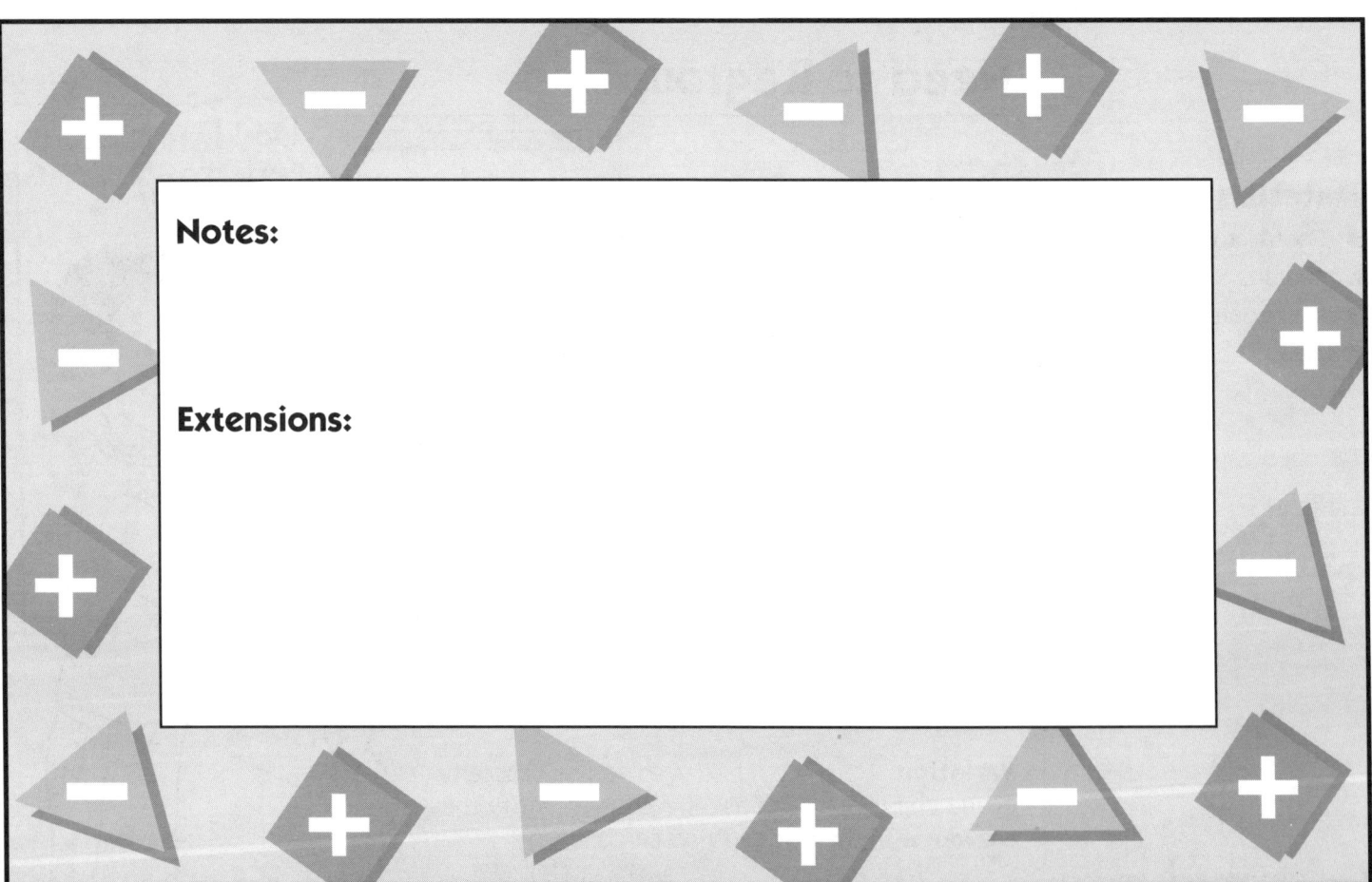

Notes:

Extensions:

Ten in a Bag

Activity 22

Materials
- beans
- small resealable bags
- chalkboard
- chalk

Procedure
1. Give each pair of students five to nine bags and ask them to place ten beans in each bag. Be sure additional beans are available to use for ones.
2. Write a subtraction problem on the board.
3. Have students model and solve the problem with their tens and ones manipulatives, trading in a bag for ten individual beans if necessary.
4. Continue as described with other problems.

Addition and Subtraction

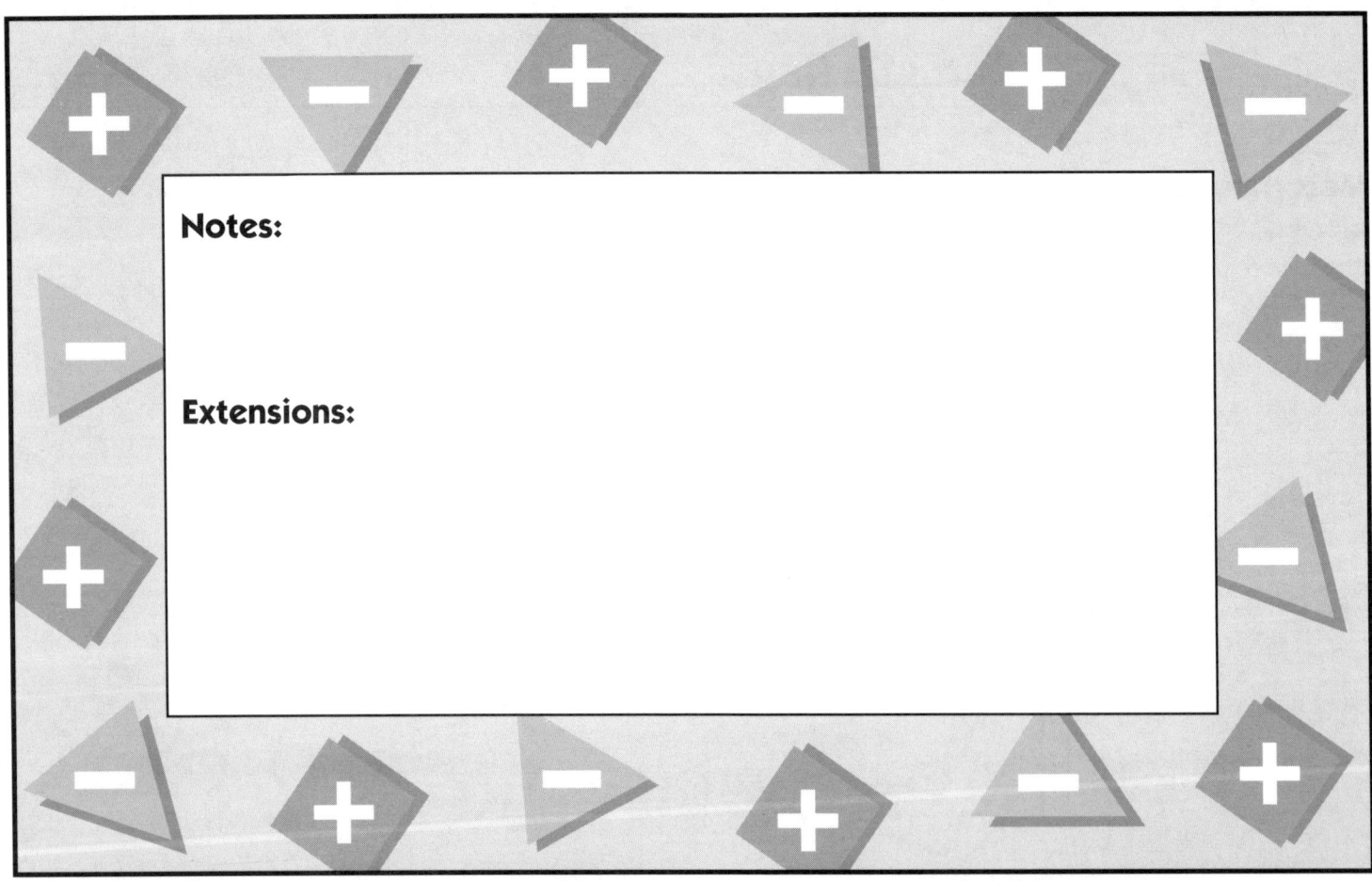

Notes:

Extensions:

Pick Three

Activity 23

Materials
- sets of number cards (labeled 1–9)
- tens and ones manipulatives (Base Ten Blocks, beansticks and beans, linking cubes)

Procedure
1. Have students work in pairs. Give each pair a set of number cards and tens and ones manipulatives.
2. Have one student from each pair pick three number cards and arrange them to form a subtraction problem, subtracting a one-digit number from a two-digit number.
3. Ask partners to use tens and ones manipulatives to solve the problem, then rearrange the cards to form a new problem.
4. Have students trade roles and repeat the activity using three new cards.

Extension Have students pick more than three cards to make more difficult subtraction problems.

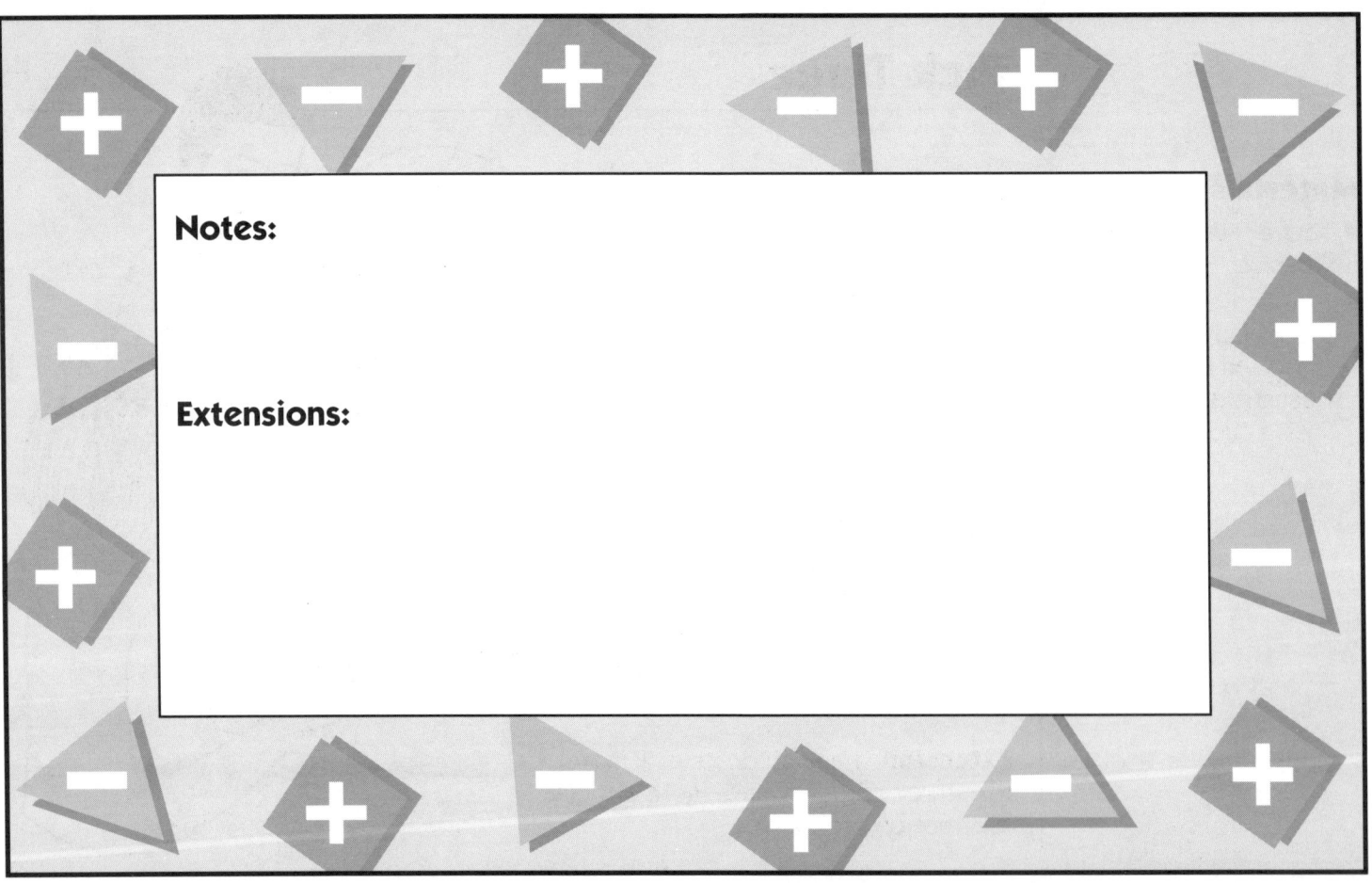

Notes:

Extensions:

Spin and Take Away

Activity 24

Materials

- tens and ones manipulatives (Base Ten Blocks, beansticks and beans, linking cubes)
- spinners (labeled 0–9)

Procedure

1. Have students work in small groups. Give each group a spinner and enough manipulatives for all students in the group.
2. Ask one student in each group to say a number from 20 to 99.
3. Have each member of the group model the number with tens and ones manipulatives.
4. Instruct another child in the group to spin the spinner and have students take away that many ones, trading one ten for ten ones if necessary.
5. Ask students to compare answers, then have another child call out a new number and continue playing.

Addition and Subtraction

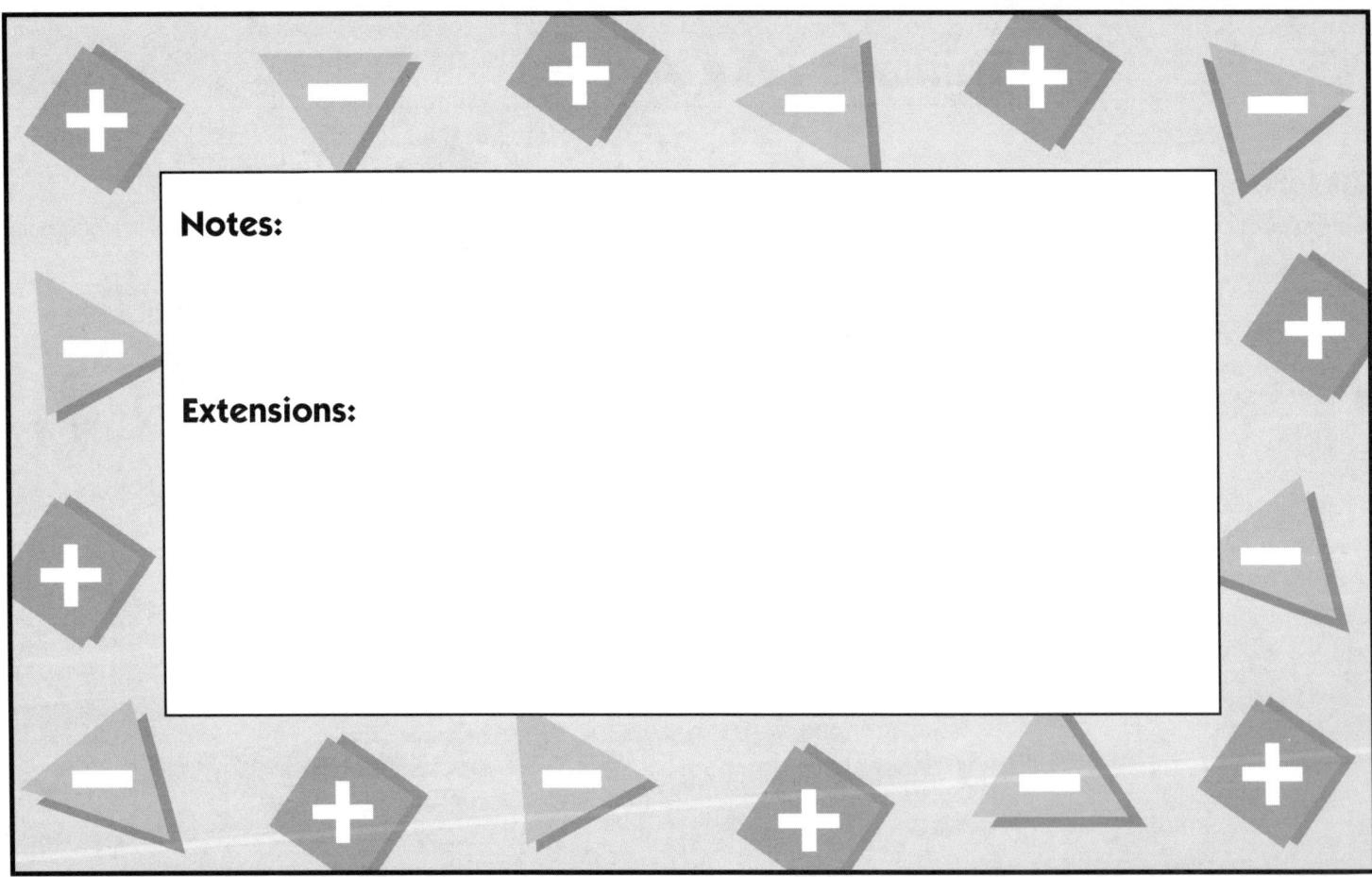

Notes:

Extensions:

Partner Subtraction

Activity 25

Materials
- dice
- paper
- pencils
- tens and ones manipulatives (Base Ten Blocks, beansticks and beans, linking cubes)

Procedure
1. Have students work in pairs. Give each pair two dice.
2. Ask Player A to roll the dice one at a time. The first number rolled stands for tens and the second number stands for ones.
3. Have Player B repeat step 1.
4. Both players decide which number is greater and write down the subtraction problem.
5. Have both players solve the problem, trading one ten for ten ones if necessary. Students can use manipulatives if desired.
6. Invite students to continue playing, repeating steps 2–5.

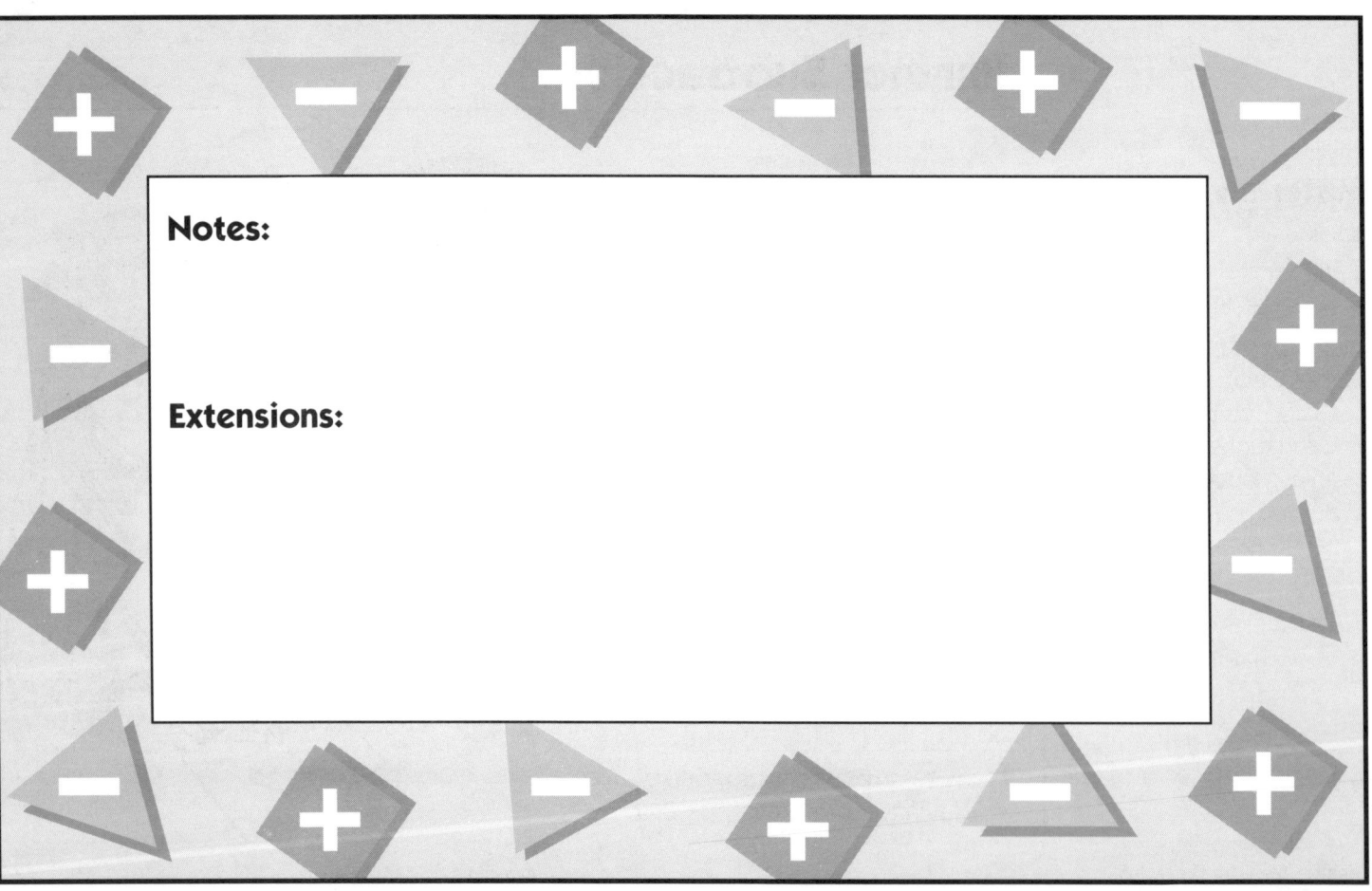

Notes:

Extensions:

Subtraction Spin

Activity 26

Materials

- tens and ones manipulatives (Base Ten Blocks, beansticks and beans, linking cubes)
- paper
- pencils
- spinners (labeled 0–9)

Procedure

1. Divide the class into small groups. Give each child five tens manipulatives. Provide a spinner and lots of ones manipulatives for each group to share.
2. Have all players write 50 at the top of their papers.
3. One child spins the spinner, subtracts that number from the five tens, and records the subtraction problem.
4. Have students take turns spinning and subtracting, trading tens and ones when necessary.
5. The first player to reach zero is the winner. (You can also set a time limit or limit the number of spins.)

Extension To practice subtraction with three-digit numbers, start the game at 200 and play to 100.

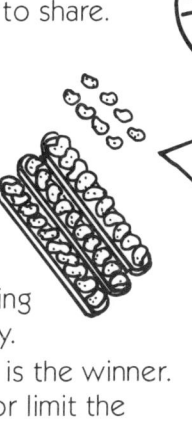

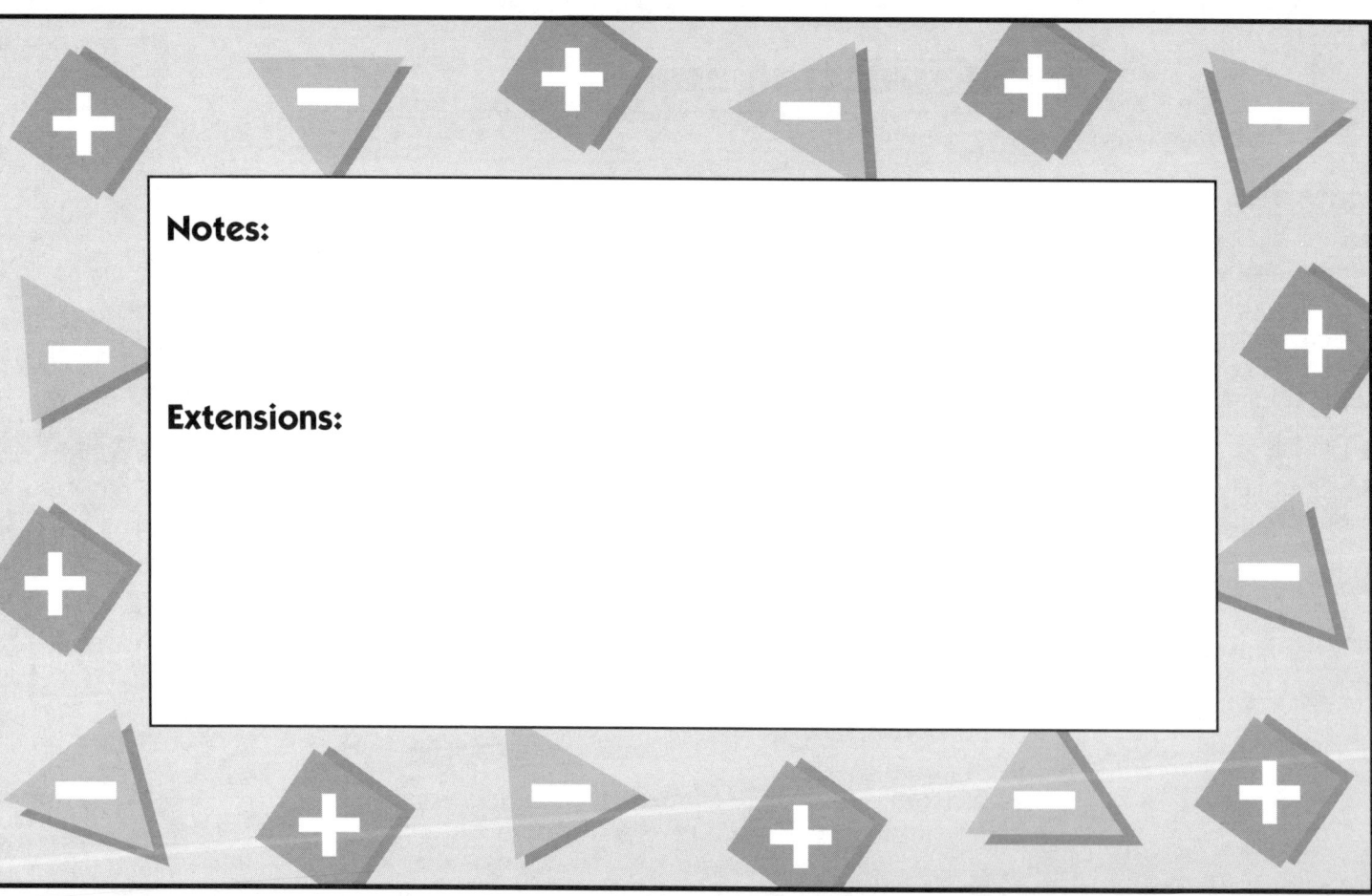

Notes:

Extensions:

Coupon Subtraction

Activity 27

Materials

- newspaper and magazine coupons
- chart paper
- markers
- paper
- pencils

Procedure

1. Collect and laminate a variety of coupons.
2. Make a chart listing the regular price of the coupon items.
3. Place materials at a center.
4. Have children compute the reduced price for each item by subtracting the value of the coupon from the regular price listed on the chart.

Extension Many grocery stores double the value of coupons. Have students find the price of double-coupon items.

Addition and Subtraction

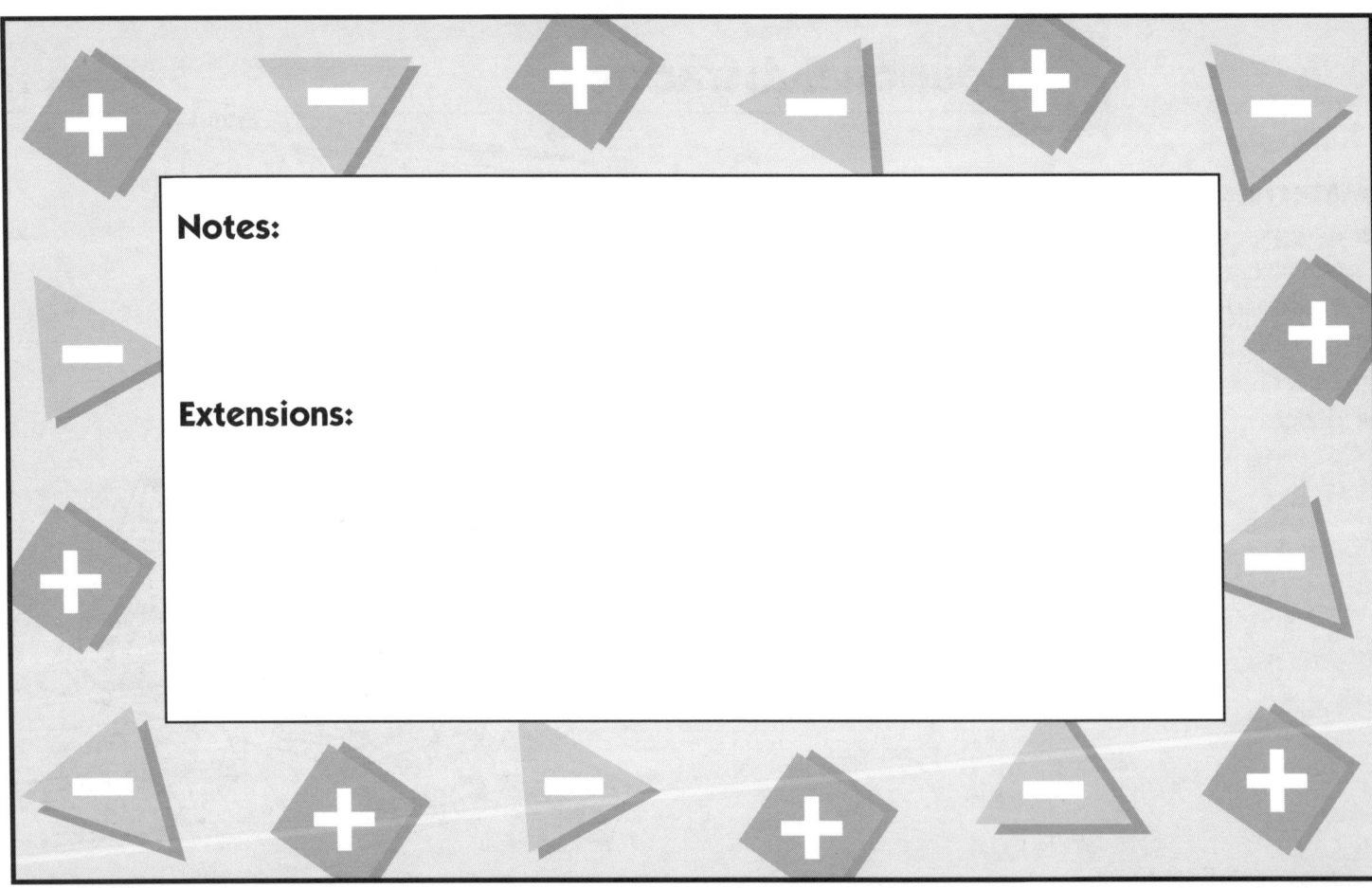

Notes:

Extensions:

Weigh and Subtract

Activity 28

Materials
- kitchen scale
- small household or classroom items to weigh
- paper
- pencils

Procedure
1. Place materials at a center.
2. Invite students to weigh two items in grams or ounces.
3. Ask students to find the difference in weight and record their work as a subtraction problem.
4. Students can take turns weighing and recording.

Variation To work on addition skills, have students select several items, weigh them, and add the total weight.

Extension Have students find the pair of items with the smallest (and greatest) difference in weight.

Addition and Subtraction

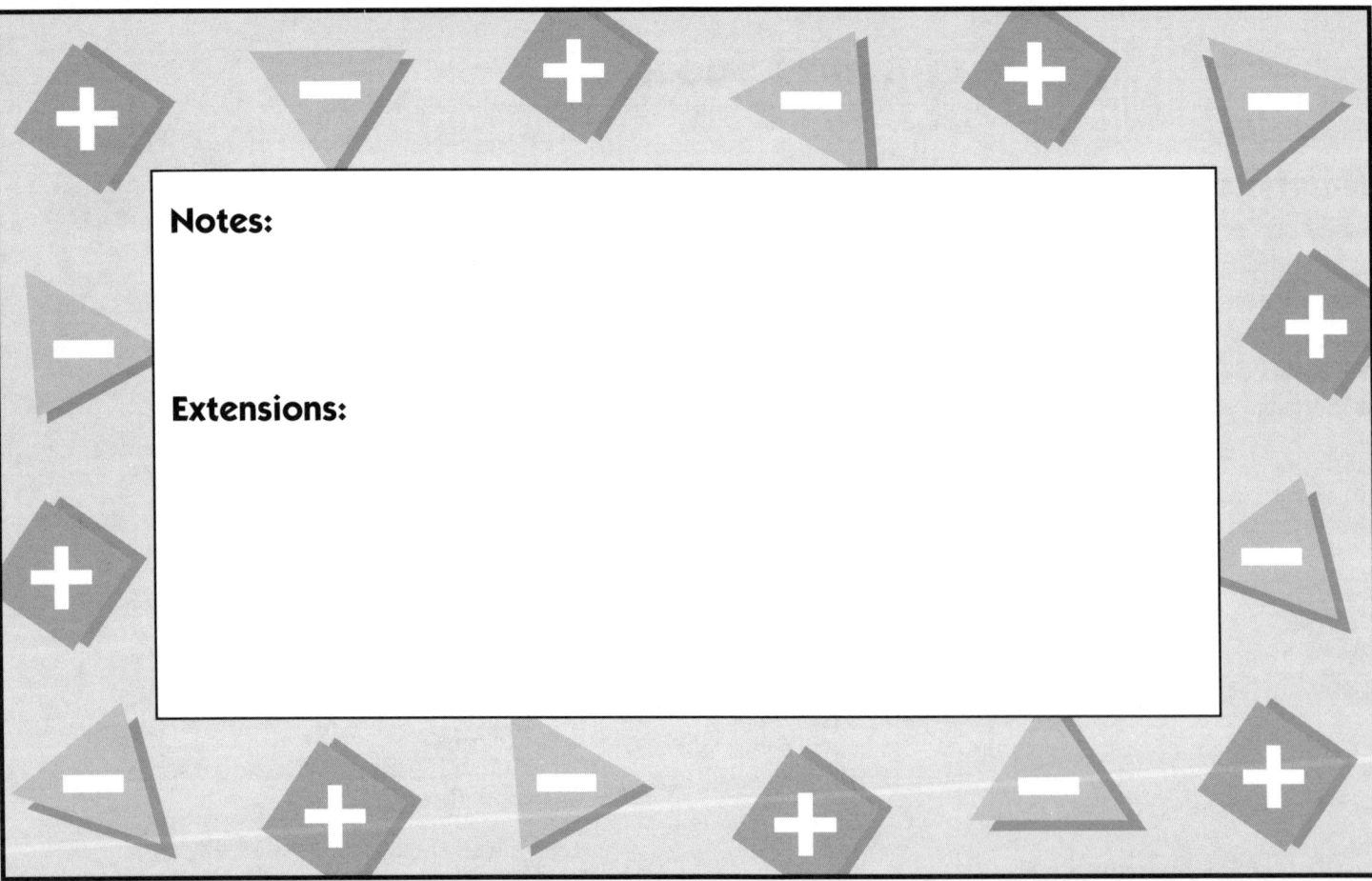

Notes:

Extensions:

Making Problems

Activity 29

Materials

- number card sets (labeled 0–9)
- paper
- pencils

Procedure

1. Divide the class into small groups of two to four students. Give each group two sets of number cards.
2. Have one player from each group shuffle all 20 cards and deal four cards to each player.
3. Players use their cards to form a subtraction problem with two-digit numbers. The goal is to make a problem with the smallest difference possible.
4. Have each player record his or her problem.
5. The player with the smallest answer gets one point. Play continues until one player has five points.

Variation Have students use the cards to form addition problems with the greatest sum possible.

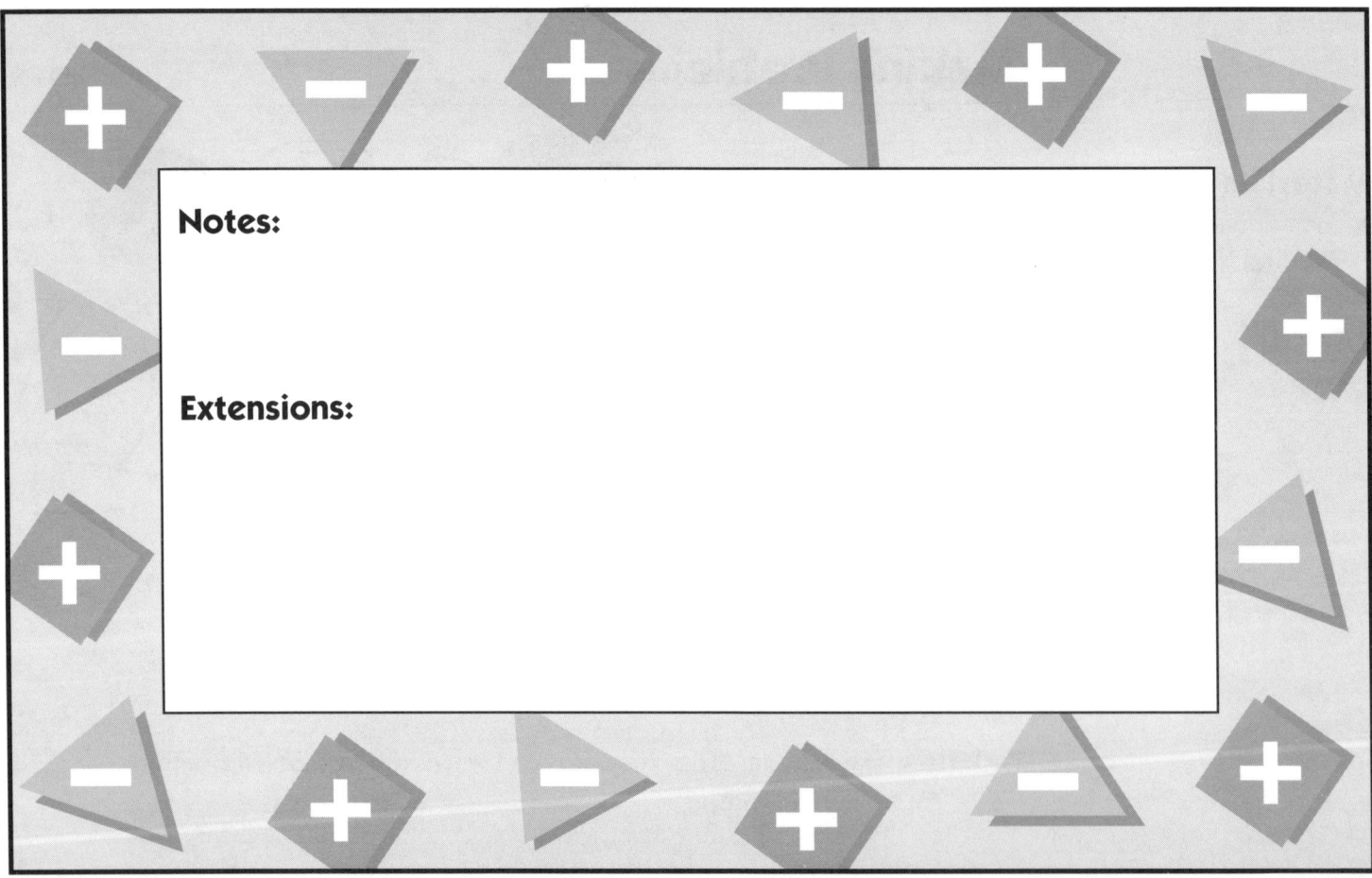

Notes:

Extensions:

Spending Ten Dollars

Activity 30

Materials

- play coins and bills
- paper
- pencils
- newspaper supermarket ads
- divided container for "bank"

Procedure

1. Place materials at a center.
2. Each student starts with a $10 bill and writes *$10.00* at the top of his or her paper.
3. Looking at the ads, have students select an item to purchase and subtract the price from $10, using coins and bills to model the subtraction. If regrouping is necessary, have students exchange bills and coins at the "bank."
4. Ask students to record the subtraction equation on paper, listing the item purchased.
5. Students continue choosing items, subtracting, and recording until they have spent their $10 (or as close as possible).

Variation Use toy store ads and give students $100 to spend.

Addition and Subtraction

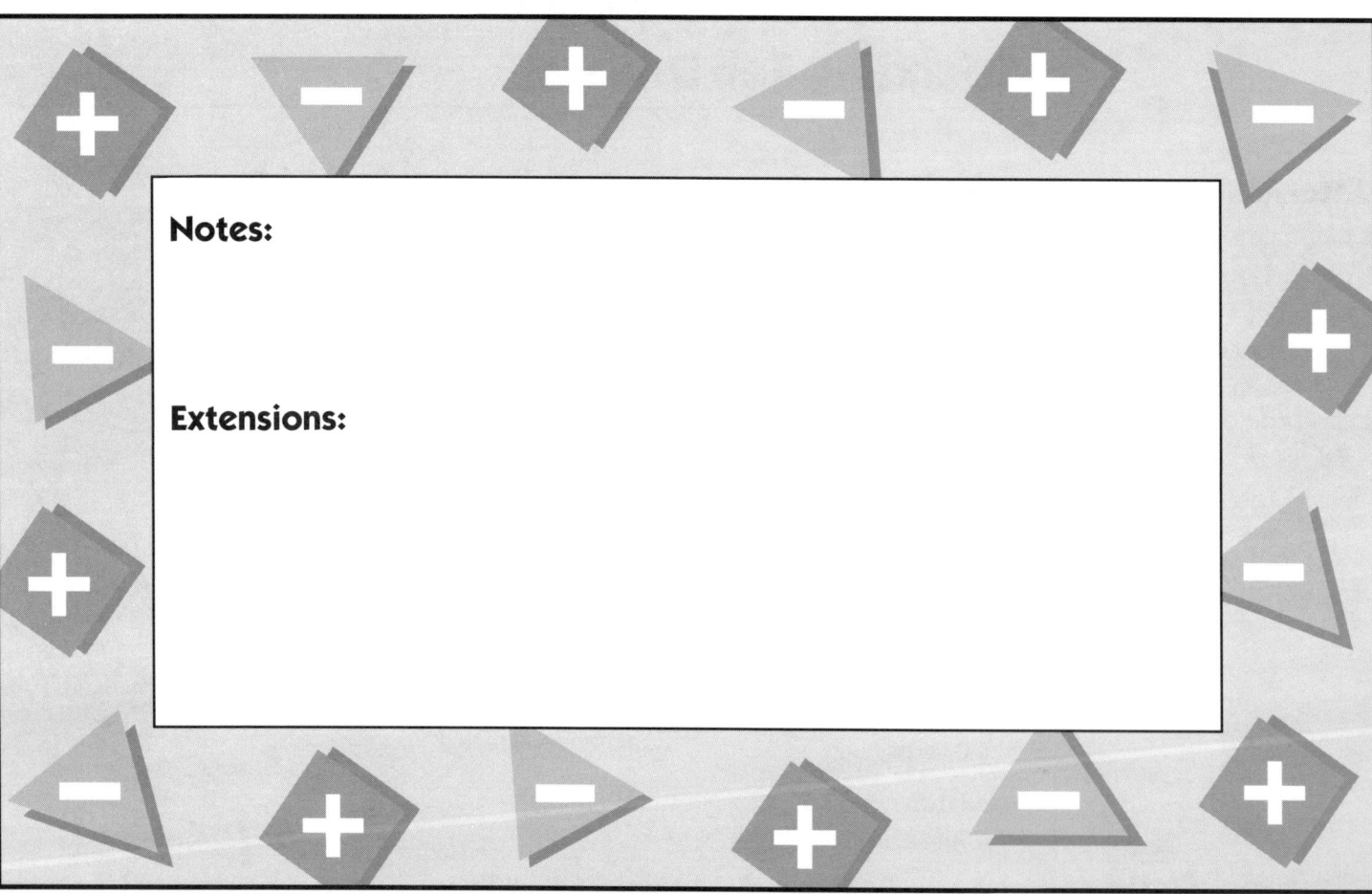

Notes:

Extensions:

Addition and Subtraction Toss

Activity 31

Materials

- 2-color counters
- small paper cups
- paper
- pencils

Procedure

1. Give each pair of students a cup and 10–20 counters.
2. Tell students how many counters to use. For example, if you want them to explore facts for 15, have them place 15 counters in the cup.
3. Have students shake and toss the counters. One student from each pair writes an addition problem and the other student writes a subtraction problem to match how the counters land. For example, if there are six red counters and nine yellow, they might write $6 + 9 = 15$ and $15 - 6 = 9$.
4. Play continues with students taking turns recording addition and subtraction problems.

Extension Have students write two addition and two subtraction problems for each roll.

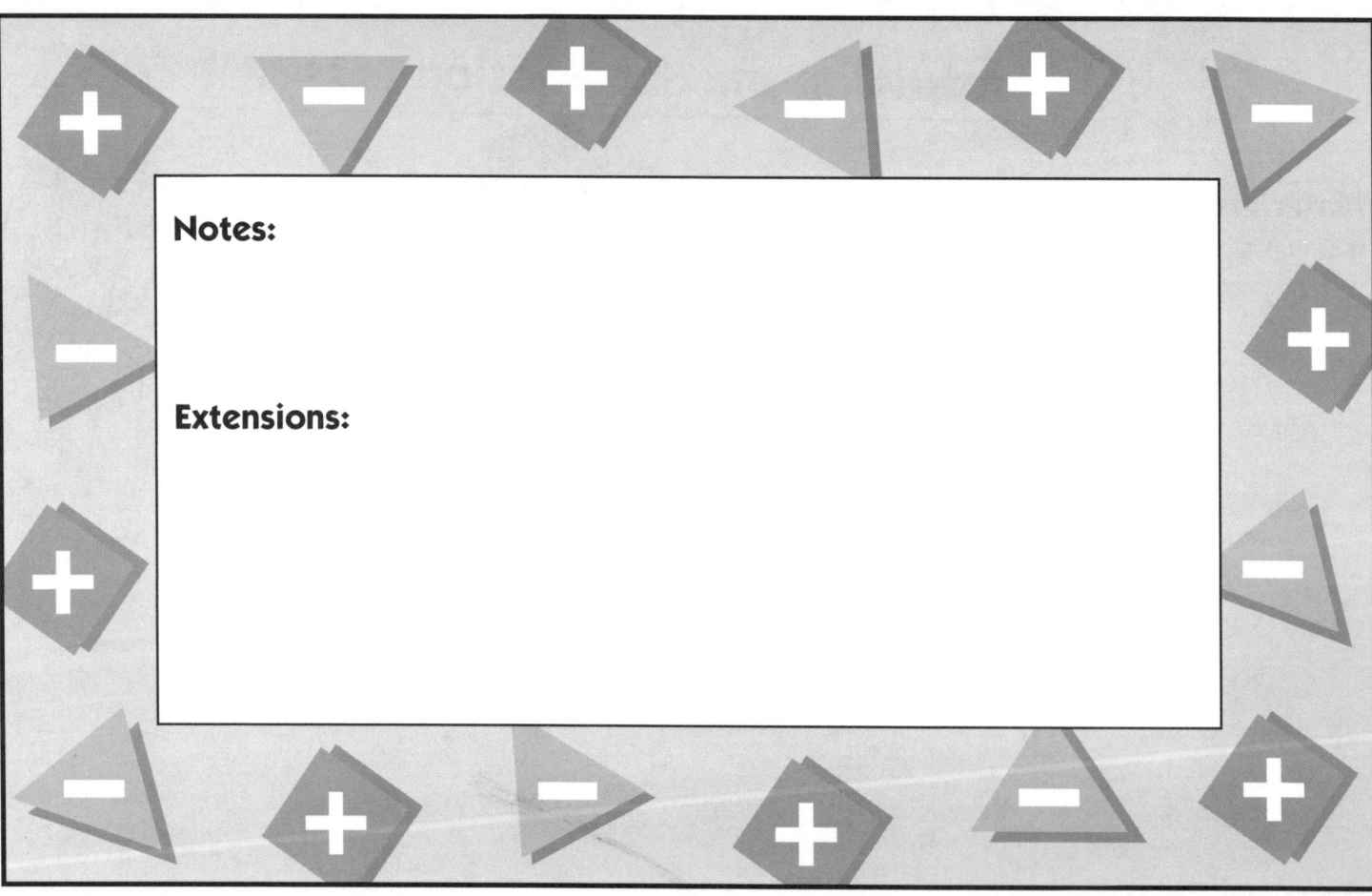
Notes:

Extensions:

Calendar Addition and Subtraction

Activity 32

Materials
- classroom calendar
- chalkboard
- chalk

Procedure
1. Point out the day's date on the calendar. Ask students to think of two numbers that add up to the date. For example, if it's March 16, a child might say, "Eight and eight." Have the child write the equation $8 + 8 = 16$ on the chalkboard.
2. Ask for other volunteers to write different addition equations for 16.
3. Then have students think of subtraction equations using 16 as the minuend, such as $16 - 9 = 7$.

Extension Challenge students to think of several subtraction problems with answers that equal the day's date. For example, $45 - 15 =$ ___ and $90 - 60 =$ ___ could both be equations for *April 30*.

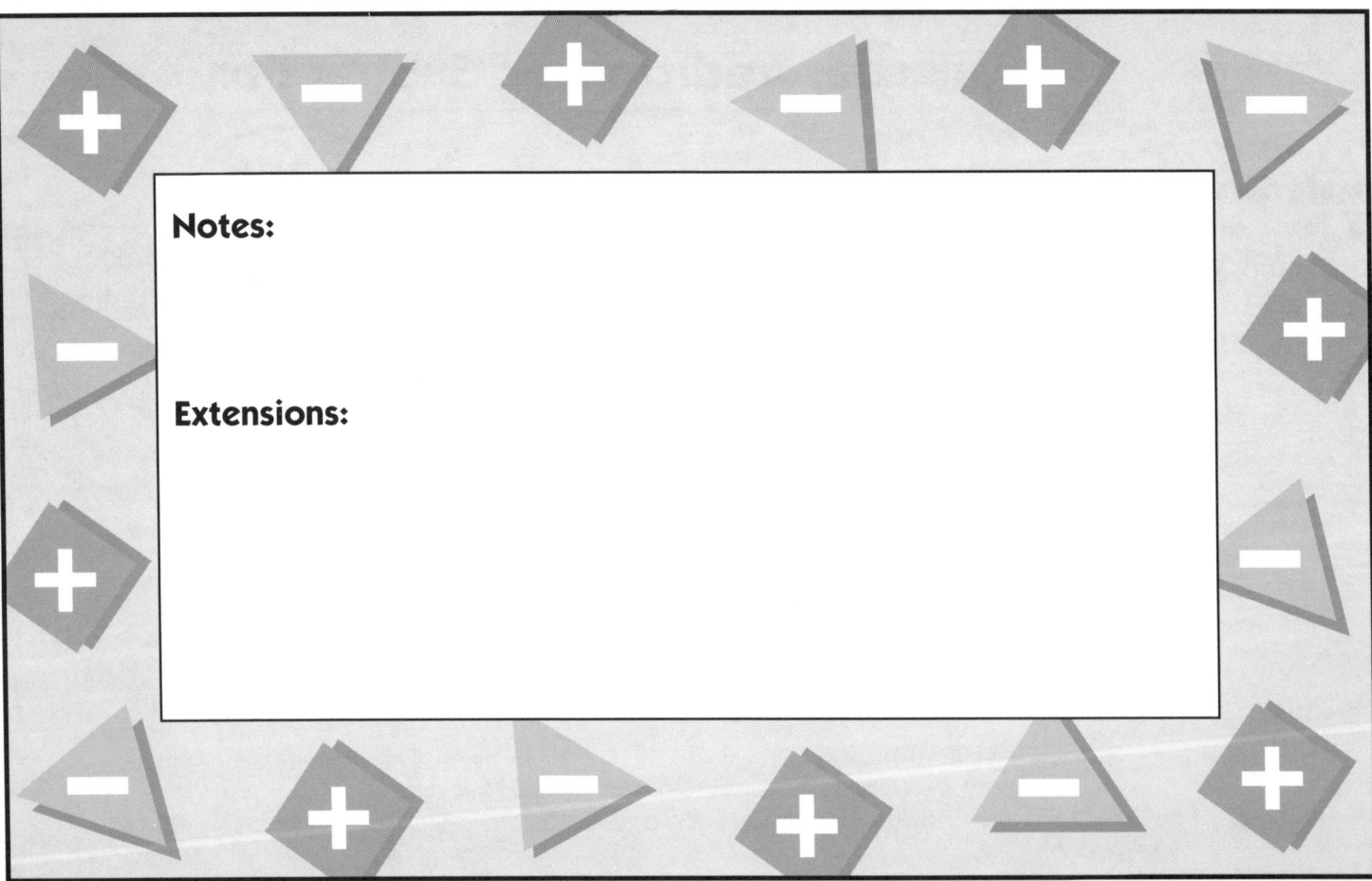

Notes:

Extensions:

Math Riddles

Activity 33

Materials
- index cards
- paper
- pencils

Procedure
1. Give each student an index card.
2. Ask students to think of two numbers and write an addition clue and subtraction clue about the two numbers on the index card.
3. Have students exchange cards and solve each other's riddles.

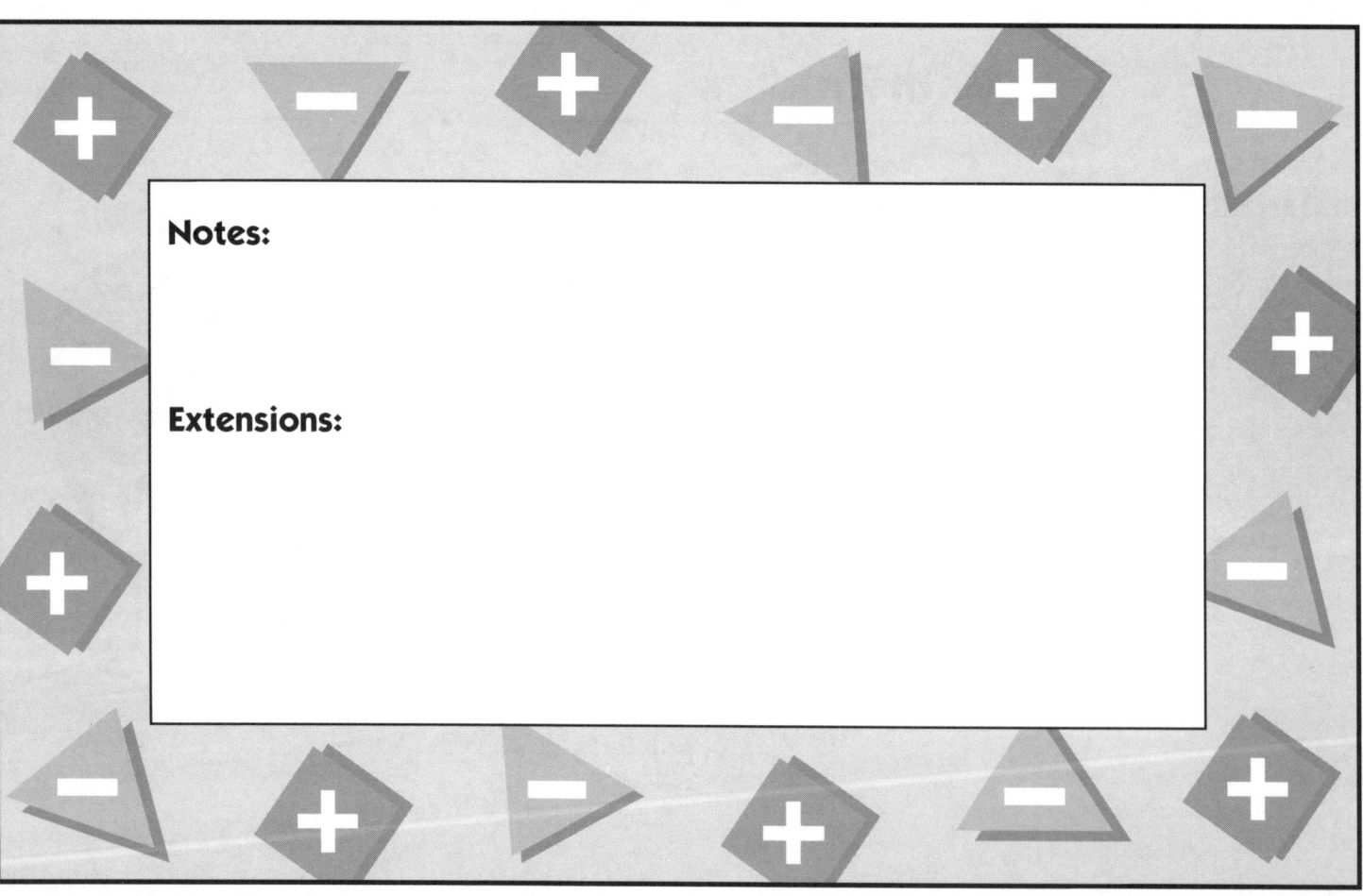

Notes:

Extensions:

Add or Subtract

Activity 34

Materials

- index cards
- markers
- flat counters marked "+" on one side and "–" on the other
- paper
- pencils

Procedure

1. Write a two- or three-digit number on each index card. Give each pair of students 10–15 cards and a counter.
2. Have students place the number cards face down.
3. Ask Player A to select two cards and flip the counter.
4. Player B must solve the indicated addition or subtraction problem.
5. Have Player A check the answer. If correct, one point is awarded.
6. Have players switch roles and continue playing, shuffling the cards when necessary.
7. Set a designated time limit. The player with the most points at the end of the play period is the winner.

Addition and Subtraction

Notes:

Extensions:

Name the Number

Activity 35

Materials

- number cards (labeled *0–20* or *0–30*)
- markers
- tape
- paper
- pencils

Procedure

1. Tape a number card on the back of each child.
2. Tell students to go around the room asking questions about their numbers. Questions must be related to addition or subtraction and must have a "yes" or "no" answer. (For example, "Is my number greater than 8 + 7?") Encourage students to keep notes about questions and responses.
3. When the number is guessed, each student removes the number but can still participate in the game by answering questions.

Notes:

Extensions:

Parent Letter

Dear Parents,

In our mathematics program, we are working on adding and subtracting large numbers. A variety of materials and activities will be used to make this unit fun and challenging for students. If you would be willing to purchase an inexpensive consumable item for these activities, please sign and return the bottom portion of this letter. I will then return the form to you stating what item to purchase and the date it is needed. Thank you!

Sincerely, _____

_____

Yes, I would like to contribute an item for addition and subtraction activities.

_____
(signature)

Please send in _____ by _____.
Thank you very much!

Tens and Ones Mat

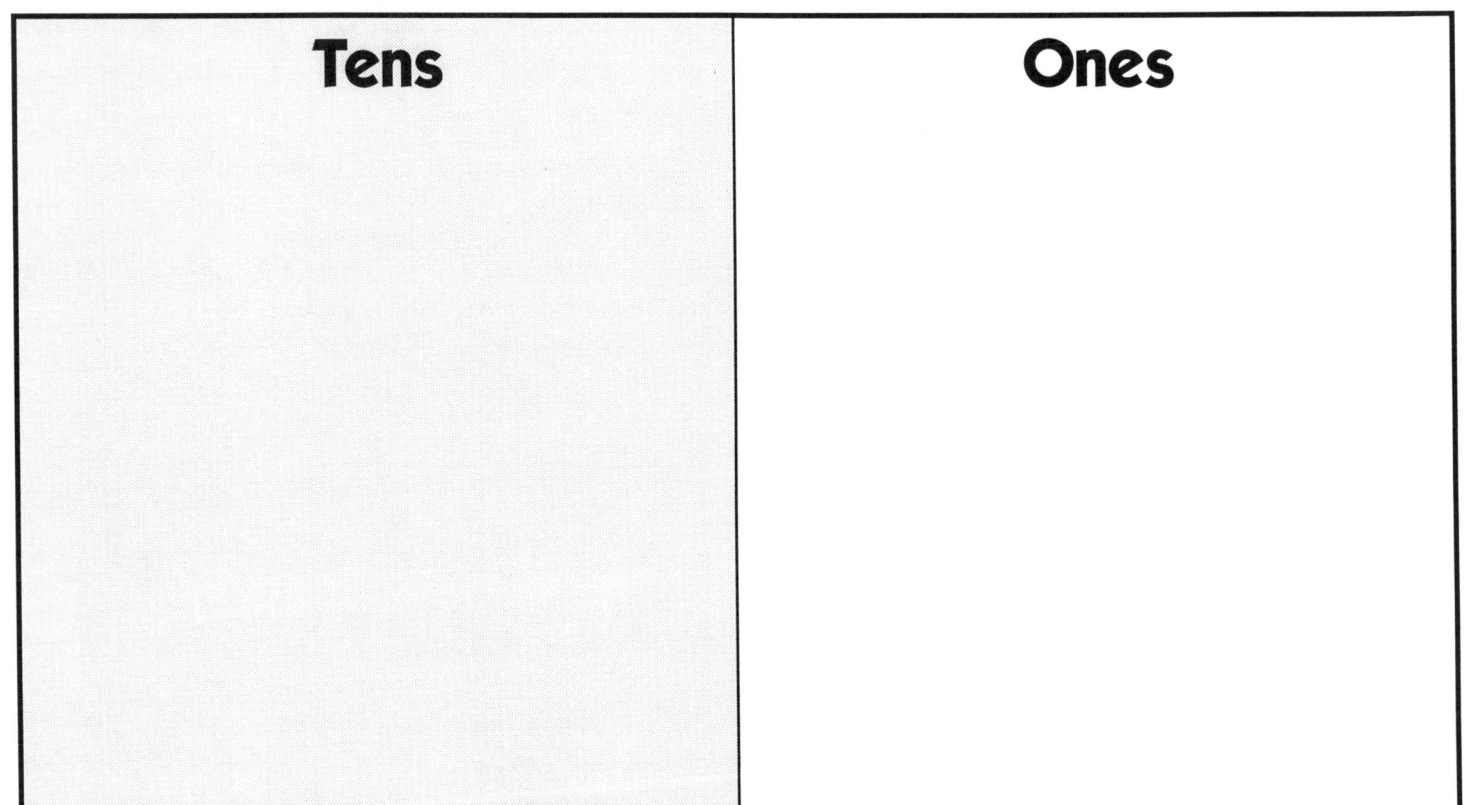

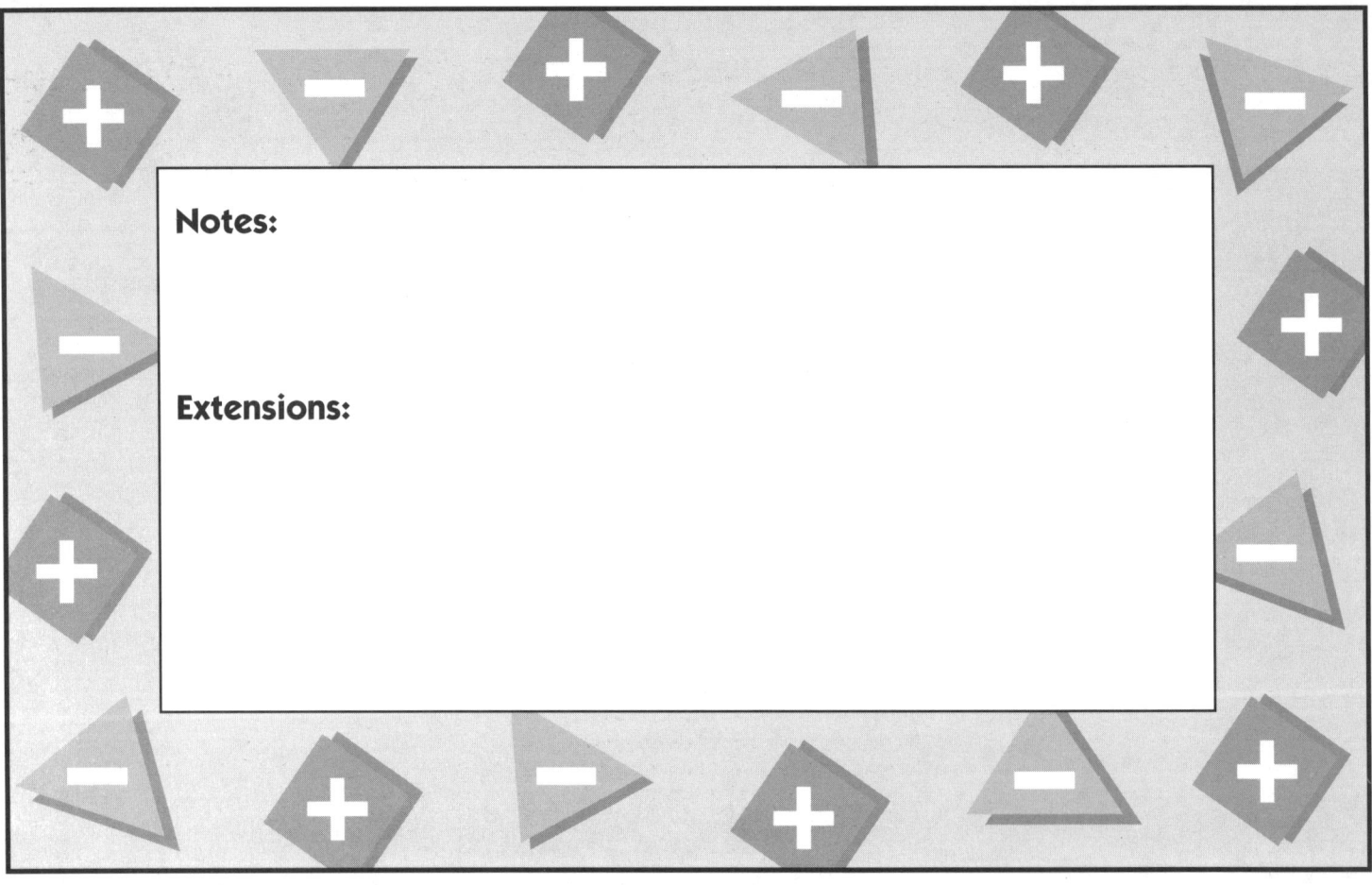

Notes:

Extensions: